Steamships across the Pacific

Crossing Seas

Editors: Henry Yu (University of British Columbia) and Elizabeth Sinn (University of Hong Kong)

The Crossing Seas series brings together books that investigate Chinese migration from the migrants' perspective. As migrants traveled from one destination to another throughout their lifetimes, they created and maintained layers of different networks. Along the way these migrants also dispersed, recreated, and adapted their cultural practices. To study these different networks, the series publishes books in disciplines such as history, women's studies, geography, cultural anthropology, and archaeology and prominently features publications informed by interdisciplinary approaches that focus on multiple aspects of the migration processes.

Books in the series:

Chinese Diaspora Charity and the Cantonese Pacific, 1850–1949
Edited by John Fitzgerald and Hon-ming Yip

Heritage and History in the China–Australia Migration Corridor
Edited by Denis Byrne, Ien Ang, and Phillip Mar

Locating Chinese Women: Historical Mobility between China and Australia
Edited by Kate Bagnall and Julia T. Martínez

Returning Home with Glory: Chinese Villagers around the Pacific, 1849 to 1949
Michael Williams

Searching for Sweetness: Women's Mobile Lives in China and Lesotho
Sarah Hanisch

Steamships across the Pacific: Maritime Journeys between Mexico, China, and Japan, 1867–1914
Ruth Mandujano López

Steamships across the Pacific

Maritime Journeys between Mexico, China, and Japan, 1867–1914

Ruth Mandujano López

Hong Kong University Press
The University of Hong Kong
Pok Fu Lam Road
Hong Kong
https://hkupress.hku.hk

© 2025 Hong Kong University Press

ISBN 978-988-8876-76-1 (*Hardback*)

All rights reserved. No portion of this publication may be reproduced or transmitted in any form or by any means, electronic or mechanical, including photocopying, recording, or any information storage or retrieval system, without prior permission in writing from the publisher.

British Library Cataloguing-in-Publication Data
A catalogue record for this book is available from the British Library.

Digitally printed

To Marty

To my mom Ruth and my dad Jesús

To Bill

To Juda

To Rosemarie

Contents

List of Maps	viii
Acknowledgments	ix
Introduction	1
1. *San Pedro*, 1565 / *Colorado*, 1867: From Sail to Steam	15
2. *Vasco de Gama*, 1874: A Space Odyssey	40
3. *Mount Lebanon*, 1884: Navigating in Britain's Diplomatic Waters	65
4. *Gaelic*, 1897: The Japanese Colonization Project in Mexico	87
5. *Suisang*, 1908: Double Vision—Chinese Migrants and the Body of the Nation	110
6. *Ancon*, 1914: Revolutions and the End of the Porfirian Transpacific System	139
Conclusion	169
Bibliography	173

List of Maps

Map 1: Vicente E. Merino, *Ruta de las flotas de España para América y la Nao de América para China* [Route followed by the Spanish fleets to America and from America's Nao to China]. 15

Map 2: Robert B. Gorsuch, *Líneas de ferrocarriles en los Estados Unidos del Norte y México* [Railway lines in the United States and Mexico]. 40

Map 3: *Vías de comunicación marítimas y terrestres* [Maritime and land routes]. 65

Map 4: M. Jaimes, *Plano del lote que pertenece a la colonia Japonesa en el Soconusco* [Map of the land belonging to the Japanese colony in Soconusco]. 87

Map 5: Alfredo A. y Carlos Vega Schiafino Jiménez, *Carta general de vías y comunicaciones de los Estados Unidos Mexicanos* [General map of Mexican roads and communications]. 110

Map 6: Francisco A. Calderón, *Carta postal y de vías de comunicación de los Estados Unidos Mexicanos* [Mexican postal chart and communication routes]. 139

Acknowledgments

"I feel lost in the middle of the Pacific without knowing where to start rowing!"
"Don't row! Fire up a steam engine! Seriously, just start and stick with it, you'll build momentum."

—Conversation of the author with Bill French

Like many of the transoceanic travels described in these pages, the end of this long book journey feels more like a shipwreck than a landing. If I reached port, it was only thanks to innumerable people who helped the project and helped me remain afloat during the emotionally exhausting and physically taxing writing process. Unfortunately, I don't have enough space nor the names of many of them as they include the strangers who smiled and shared instances of comradery, the cooks of the meals I ate, the musicians whose songs lifted my spirits, and many others. The following are only some of the most important and direct contributors. I wish I could thank everyone who supported me in one way or another over the past years; I apologize for all the unintentional omissions.

This journey started while I was doing my PhD at the University of British Columbia, where I had the good fortune and privilege to work with the most clever, dedicated, and generous faculty on my dissertation committee. Despite overstuffed agendas, Alejandra Bronfman, Bill French, and Henry Yu, later joined by Evelyn Hu-DeHart as an external, always found the time, offering encouragement and solid advice whenever I needed them. My talented classmates and friends Gabriela Aceves Sepúlveda and Laura Madokoro offered a much-needed sorority throughout the full array of happy and challenging times.

Three institutions contributed financially to my PhD research and manuscript, which forms the basis of this book: in Mexico, the Consejo Nacional de Ciencia y Tecnología (CONACYT), and, in Canada, the Social Sciences and Humanities Research Council of Canada (SSHRC), and the University of British Columbia, particularly through the Faculty of Graduate Studies and the History Department.

A previous version of Chapter 3, "*Mount Lebanon*, 1884: Navigating in Britain's Diplomatic Waters," was published in the *International Journal of Maritime History*.

I am indebted to all the editors, in particular to Rights Administrator Craig Myles, for granting permission to reuse it in this book. I discussed data from Chapters 1 and 3 in the journal *Retos Internacionales* and from sections related to maritime companies and travelers, notably from Chapter 6, in my contribution to the book *Intercambios, actores, enfoques. Pasajes de la historia latinoamericana en una perspectiva global*. I am also indebted to the founder and director Gabriel Morelos Borja of the former, and the editor Aarón Grageda of the latter, for authorizing its inclusion here.

Douglas College, where I have worked since 2007, offered me an Educational Leave which provided the funds and the time to write the last chapters. I particularly thank my Dean, Manuela Costantino, for her continuous support; all my Modern Languages colleagues for their encouragement and for taking on the extra load that my leave put on them; the librarians who helped with interlibrary loans; the Student Research Assistant program that enabled Maria Luiza Zenobio de Assumpção Villar Laufhütte to help me gather the bibliography; and all the students who have shared their laughter and thoughts with me in the classroom.

No history book would come to fruition without the support of all the dedicated and patient staff in the archives and libraries as well as the people willing to share their family or personal memories, books, and documents. In particular, I am grateful to everyone in the Mapoteca "Manuel Orozco y Berra" in Mexico City, who facilitated access to maps for the book and made the process easy and cost-free. I recommend to everyone that they visit their beautiful facilities or browse their extensive and well-organized online catalogue. I also thank my MA supervisor and friend, Mónica Toussaint Ribot, who shared her research and connections whenever I had questions, as well as my dearest friend Judith Espinosa Rayas, who lent me books from her extensive collection and even bought additional materials that I needed to complete the project on time.

I also want to thank the team at Hong Kong University Press who have supported me throughout the process, in particular Yasmine Hung, who has been my direct contact and answered all my questions with patience and kindness since the beginning. I am also indebted to the two anonymous readers, who thoroughly examined the manuscript and provided interesting suggestions to improve it.

The list of family and friends with whom I have shared the past years and who have, whether explicitly or inadvertently, played a part in this project is long. In Vancouver, my in-laws, our inner gang of friends (you know who you are), our football teams (Soccerinos, Academix Anonymous), my Canto Vivo choir, and my Polymer Dance class helped transform long winters and writing hours into joyful times. In Mexico, I thank my family, the Corner, and the rest of my friends, who have formed an integral part of my life for as long as I can remember. In particular, I would like to express appreciation to my aunts Gladys and Velly and my uncle Carlos for sharing family stories related to the book's topic and, overall, my mother

Acknowledgments

Ruth, for her eternal love and patience, and for her delicious meals while I was writing part of the manuscript.

There are three key people without whom I would not have been able to finish this book. Not only did they read the manuscript and suggest clever edits to improve it, but they also gave me the confidence and encouragement I needed to get to the finish line. Henry Yu has sparked my curiosity and made me rethink migration since the first PhD course I took with him. He has invited me to participate in many interesting projects throughout the years, including an unforgettable trip to Yuquot, on the Canadian Pacific coast, which helped shape some of the ideas discussed in the Introduction, as well as the invitation to contribute to the Crossing Seas series. My immense gratitude goes to Bill French, who went far beyond his responsibility as PhD supervisor. Bill became both mentor and close family friend after our first meeting during his unforgettable Oaxacan seminar in 2006. Whether it was to clarify a concept or to discuss the latest film, Mexican politics, or my most pressing personal and academic doubts, Bill has been there, constantly supporting my initiatives and sharing his home and delicious meals whenever our paths coincided. With humor, patience, intellectual rigor, and kindness, he helped me navigate the writing of this book. Finally, and foremost, I thank my partner in life, Marty McLennan, for the love, the patience, the emotional and intellectual support, the care when I was sick or down, the delicious flavors, the jokes, the cleaning, the help in the archives, the editing of this and many other manuscripts, the interesting discussions, the amazing discoveries, the trips, the songs, the films, the tales, the laughs, the dreams. Thank you for all of these and for sharing the most wonderful twenty-five years together.

Introduction

A routine health inspection of a steamship on the morning of May 15, 1908, turned out to be anything but ordinary. When Dr. Valenzuela, Health Delegate of Salina Cruz, entrusted with performing a general examination of passengers docking at the southern Mexican port, boarded *Suisang*, a steamer of the China Commercial Steamship Company (CCSC) that plied the waters between Hong Kong and Salina Cruz on a monthly basis, he found 10 second-class passengers and close to 400 of those in steerage to be infected with trachoma, a contagious eye disease. Unsure of how to proceed, he asked federal authorities for instructions. They told him to "suggest" to the second-class passengers that they isolate themselves in barracks, as they had no authority to compel them to take such a step; as for those in steerage, they were given no choice but to remain on board until further notice. First- and second-class passengers not presenting symptoms were permitted to land without additional procedures; those in steerage were forced to undergo a ten-day quarantine in the barracks, where they were made to take baths and their luggage was fumigated.[1]

The CCSC had signed a contract with the Mexican government in 1903 to establish a regular steamship service between Hong Kong or other ports of China, Japan, and the United States and the Mexican port of Manzanillo.[2] The company eventually added Salina Cruz as one of its terminals; it brought, on average, 480 monthly passengers to Mexico. While in previous inspections sick passengers had been put in quarantine, this was the first time that such a numerous contingent had been found to be ill.[3] After the quarantine, 111 steerage passengers deemed in good health were given their liberty, but the port's authorities ordered 28 who had developed symptoms to re-embark; a total of 413 ended up locked in the *Suisang*

---

1. Public Records Office–Hong Kong (PRO-HK), C.O.129/352, 342–44 and C.O.129/378, 102.
2. "Contrato," *Diario Oficial*, February 1903, 677.
3. Before *Suisang*'s arrival in May 1908, the largest number of CCSC passengers rejected at Salina Cruz had been forty-eight. PRO-HK, C.O.129/349, 312.

without permission to land.⁴ All the second-class passengers were allowed to leave even though they were still suffering from trachoma, four of them "in an advanced stage and very serious form."⁵ CCSC representatives filed a complaint arguing that, according to the ship's surgeon, the disease was not trachoma but a simple conjunctivitis that was not contagious. In their view, it was unlawful to forbid the remaining steerage passengers to disembark.⁶ This triggered a long diplomatic dispute involving Mexican, Chinese, and British authorities and passengers, with catastrophic consequences for many of those involved.⁷

Until recently, it was widely accepted that links between Mexican and Asian ports had ceased or, at best, become irrelevant as of 1815, when the famous 250-year-long route between Manila in the Philippines and Acapulco in Mexico was shut down as the latter became independent from Spain.⁸ But as the *Suisang* and the CCSC case, as well as others analyzed in this book, illustrate, such connections existed and mattered, not only for those directly involved, but also as part of broader local, regional, and global interactions and exchanges. The central premise that guides this book is that during the second half of the nineteenth and the beginning of the twentieth centuries, when steamships became the main international mode of transportation for people and merchandise, they also became sites of contention where shifting power relations between individuals, groups, communities, nations, and empires were configured, negotiated, exerted, and reimagined. By studying the maritime connections and the specific steamers, companies, and people that circulated between Mexican and Asian ports, my aim is to contribute to a more complex picture of the individual, regional, and global transformations happening around the globe, and particularly in the Pacific basin, when industrial changes hit maritime transportation.

Mexican–Asian Connections in Historical Studies

I spent my childhood summers in my grandparents' home, on the Pacific coast of Chiapas, hearing family stories that often included old acquaintances with foreign heritage.⁹ Coming from Mexico City, with its millions of inhabitants, it surprised

---

4. PRO-HK, C.O.129/352, 345.
5. Only one suffering from mumps remained isolated for a few more days and then was released on June 1. PRO-HK, C.O.129/352, 345–47.
6. PRO-HK, C.O.129/378, 105–6.
7. The case will be thoroughly analyzed in Chapter 5.
8. Even though isolated galleons traveled to Mexican ports up to the 1820s, most researchers agree on 1815 as the official end date for the route. See Peter Gordon and Juan José Morales, *La plata y el Pacífico: China, Hispanoamérica y el nacimiento de la globalización, 1565–1815* (Madrid: Siruela, 2022), 98; Vera Valdés Lakowsky, *Vinculaciones sino-mexicanas: albores y testimonios, 1874–1899* (Mexico City: UNAM, 1981), 62.
9. As will become evident in subsequent chapters, Mexican ports had direct and regular steamship connections with Asia only with Hong Kong and Yokohama during the period analyzed in this book, so the historiographical review focuses on those ports and Chinese and Japanese travelers and migrants. This is not to suggest that other ports and nationalities do not matter, but they are not the focus of this book.

me that such a small place would have more diversity than what I experienced in the "big city." But the explanation was simple: Tonalá, my mother's hometown, connected to the railway system since the beginning of the twentieth century,[10] lies between two of the largest Mexican Pacific ports, Tapachula to the east and Salina Cruz (where the CCSC vessels landed) to the west.[11] My grandfather worked as a coal stoker and assistant for the trains that connected the two ports, and my grandmother and her twelve children peddled the products that he brought from Tapachula to supplement the family's tight income. It is therefore not surprising that they knew many people, including some with Asian heritage. In the stories I heard, part of my family referred to them as entirely part of the community, while others saw them as foreigners regardless of the fact that many were actually born on Mexican soil.

My family's ambivalent attitude toward Mexicans with Asian heritage as both insiders and outsiders reflects the kinds of debates that academics in Mexico have had for decades. On the one hand, studies on foreigners and on the country's foreign relations pioneered the research on Asians in Mexico.[12] On the other hand, many researchers have long questioned the assumed alienness of Asian heritage in a place with such a long history of regular transpacific connections. Their combined studies have convincingly shown that Asian otherness is a creation with Spanish colonial antecedents that consolidated with the Mexican Revolution, which broke out in 1910, and its nationalist ideologies based on racial improvement that centered around the notion of the "Mestizo" as the ideal (and often the only) way of being Mexican.[13] While "Mestizo" is a complex and highly contested term that continues to

---

10. Valente Molina Pérez, "Impacto económico y social del Ferrocarril Panamericano en la región de Tonalá en el siglo XX," *Revista Pueblos y Fronteras Digital* 11, no. 21 (January–June 2016): 72–74, https://doi.org/10.22201/cimsur.18704115e.2016.21.3.
11. Tapachula is the name of the city and the municipal territory, and it is also the name that my family used. The actual port, located some thirty kilometers from Tapachula city, was called San Benito during the period studied in this book and is now called Puerto Chiapas.
12. Two studies and compilations by Moisés González Navarro were pioneering in the field: *La colonización en México, 1877–1910* (Mexico City: Talleres de Impresión de Estampillas y Valores, 1960) and *Los extranjeros en México y los mexicanos en el extranjero, 1821–1970*, Vols. 1, 2 (Mexico City: El Colegio de México, 1993–1994). Subsequently, María Elena Ota Mishima published more detailed compilations on Asian migrations in *Siete migraciones japonesas en México, 1890–1978* (Mexico City: El Colegio de México, 1982) and *Destino México: Un estudio de las migraciones asiáticas a México, siglos XIX y XX* (Mexico City: El Colegio de México, 1997). Relevant works on diplomatic history with a focus on China and Japan include Mercedes de Vega et al., *Historia de las relaciones internacionales de México, 1821–2010*, Vol. 6: *Asia* (Mexico City: SRE, 2011); Carlos Uscanga, "Hacia una contextualización histórica de las relaciones diplomáticas de México y Japón," *Revista Mexicana de Política Exterior* 86 (June 2009): 67–89; Shicheng Xu, "Algunas reflexiones sobre el desarrollo de las relaciones sino-mexicanas," *Cuadernos Americanos* 121 (July–September 2007): 171–86; Jorge A. Schiavon et al., eds., *En busca de una nación soberana. Relaciones internacionales de México, siglos XIX y XX* (Mexico City: SRE, 2006); Vera Valdés Lakowsky, "México y China: del galeón de Manila al primer tratado de 1899," *Estudios de Historia Moderna y Contemporánea de México* 9 (1983): 9–19; and Valdés Lakowsky, *Vinculaciones sino-mexicanas*.
13. Two of the most recent and complete monographs on Asians in colonial Mexico are Tatiana Seijas, *Asian Slaves in Colonial Mexico: From Chinos to Indians* (New York: Cambridge University Press, 2014) and Déborah Oropeza, *La migración asiática en el virreinato de la Nueva España: un proceso de globalización, 1565–1700* (Mexico City: El Colegio de México, 2020).

be debated, in its simplest form it acknowledges Spanish and First Nations (referred to as Indígenas in the Mexican context) mixed ancestry, often to the detriment of any other heritage.[14] In this context, studies on *antichinismo* or Sinophobia have been particularly fruitful in showing that Chineseness has often been portrayed as foreign by Mexican politicians or specific interest groups with the aim of creating cohesion among a heterogeneous population and enforcing developmentalist projects under their nationalist and paternalistic supervision.[15]

More recently, the study of Asian–Mexican and, more broadly, Asian–Latin American connections has been expanded by the contributions of researchers coming from Asian American, transnational, borderlands, and diaspora studies.[16] Rather than focusing on national politics as their predecessors mostly did, they have

---

14. This basic definition of Mestizo was prevalent for the period studied in this book, but the concept will be addressed more thoroughly in Chapter 3. There is an extensive literature on Mestizaje. The following are only some of the key works in English, cited chronologically: Moisés González Navarro, "Mestizaje in Mexico during the National Period," in *Race and Class in Latin America*, ed. Magnus Mörner (New York: Columbia University Press, 1970); Alan Knight, "Racism, Revolution and Indigenismo: Mexico, 1910–1940," in *The Idea of Race in Latin America, 1870–1940*, ed. Richard Graham (Austin: University of Texas Press, 1990); Peter Wade, *Race and Ethnicity in Latin America* (London: Pluto Press, 1997); Alexandra Minna Stern, "From Mestizophilia to Biotypology: Radicalization and Science in Mexico, 1920–1960," in *Race and Nation in Modern Latin America*, ed. Nancy Appelbaum, Anne Macpherson, and Karin Alejandra Rosemblatt (Chapel Hill: University of North Carolina Press, 2003), 187–210; Ana María Alonso, "Conforming Disconformity: 'Mestizaje,' Hybridity, and the Aesthetics of Mexican Nationalism," *Cultural Anthropology* 19, no. 4 (November 2004): 459–90; Laura Gotkowitz (ed.), *Histories of Race and Racism: The Andes and Mesoamerica from Colonial Times to the Present* (Durham, NC: Duke University Press, 2011); Edward E. Telles, *Pigmentocracies: Ethnicity, Race and Color in Latin America* (Chapel Hill: University of North Carolina Press, 2014); Christina A. Sue, "Is Mexico Beyond Mestizaje? Blackness, Race Mixture, and Discrimination," *Latin American and Caribbean Ethnic Studies* 18, no. 1 (January 2023): 47–74. I thank Bill French for helping me navigate this historiography.
15. The most recent and complete study in English on *antichinismo* and the creation of Mexican identity is Jason Oliver, *Chino: Anti-Chinese Racism in Mexico, 1880–1940* (Champaign: University of Illinois Press, 2017). Earlier works focused on antichinismo in Mexico also used in this book are Charles C. Cumberland, "The Sonora Chinese and the Mexican Revolution," *The Hispanic American Historical Review* 40, no. 2 (May 1960): 191–211; Philip A. Dennis, "The AntiChinese Campaigns in Sonora, Mexico," *Ethnohistory* 26, no. 1 (Winter 1979): 65–80; Evelyn Hu-DeHart, "Racism and Anti-Chinese Persecution in Mexico," *Amerasia Journal* 9, no. 2 (1982): 1–28; Humberto Monteón González and José Luis Trueba Lara, *Chinos y antichinos en México. Documentos para su estudio* (Guadalajara: Gobierno del Estado de Jalisco, 1988); José Luis Trueba Lara, *Los chinos en Sonora: una historia olvidada* (Hermosillo: Universidad de Sonora, 1990); Knight, "Racism, Revolution, and Indigenismo; José Jorge Gómez Izquierdo, *El movimiento antichino en México, 1871-1934. Problemas del racismo y del nacionalismo durante la Revolución Mexicana* (Mexico City: INAH, 1991); Juan Puig Llano, *Entre el río Perla y el Nazas: la China decimonónica y sus braceros emigrantes, la colonia china de Torreón y la matanza de 1911* (Mexico City: CNCA, 1992); Gerardo Rénique, "Race, Region, and Nation: Sonora's Anti-Chinese Racism and Mexico's Postrevolutionary Nationalism, 1920s–1930s," in *Race and Nation in Modern Latin America*, ed. Nancy P. Appelbaum, Anne S. Macpherson, and Karin Alejandra Rosemblatt (Chapel Hill: University of North Carolina Press, 2003), 211–36; José Luis Chong, "Hijo de un país poderoso. La inmigración china a América (1850–1950)," *Diacronías. Revista de Divulgación Histórica* 1, no. 1 (February 2008): 55–64.
16. Since this book focuses on Chinese and Japanese links with Mexico, my historiographical research has also focused on them and not on other ethnicities. The following two historiographical essays have done an excellent job at discussing the literature on Asian migrations to Latin America: Tatiana Seijas, "Asian Migrations to Latin America in the Pacific World, 16th–19th centuries," *History Compass* 14 (2016): 573–81, and Jian Gao, "Chinese Migration to Latin America: From Colonial to Contemporary Era," *History Compass* 19, no. 9 (2021): 1–13. I thank Fredy González for referring me to these articles.

introduced a broader perspective that places each case study within a transnational, Latin American, US–Mexico borderlands, hemispheric, Pacific, and/or global context, allowing for more comprehensive explanations and for people's voices to be taken more into consideration. A key pioneer has been Evelyn Hu-DeHart with her early studies on Chinese merchants in northern Mexico, followed by various other topics such as Chinese labor migrants in the Americas, Chinese diasporas in Latin America, comparative studies in race and ethnicity, and Chinese in colonial Spain, among others.[17] Hu-DeHart, along with, subsequently, Daniel Masterson, Jeffrey Lesser, Robert Chao Romero, Erika Lee, Grace Peña Delgado, Fredy González, Elliot Young, Julia Schiavone Camacho, and others, have successfully shown that, as of the 1880s, Asians, in particular Chinese and Japanese, created interconnected transnational orbits (Romero), communities (Masterson), networks (Hu-DeHart, Peña, Young), and diasporic communities (González, Lesser) and/or citizenry (Lee, Schiavone) throughout the Pacific Rim and especially along the American hemisphere. They did so to advance their interests and, at times, to adapt to Chinese and Japanese exclusion laws imposed throughout the continent, starting with the Chinese Exclusion Act (1882) for the Chinese and the Gentleman's Agreement (1907) for the Japanese in the United States. All these studies have also determined that the transoceanic travelers in the period studied in this book were overwhelmingly male.[18]

This book has taken this historiography as a point of departure. However, it is less concerned with how migrants adapted and formed transpacific networks to

---

17. Evelyn Hu-DeHart has published an extensive scholarship. Cited here are only a few of her publications that have been used in this book, starting from the 1980s to present: "Racism and Anti-Chinese Persecution in Mexico"; "Latin America in Asia-Pacific Perspective," in *What Is In a Rim? Critical Perspectives on the Pacific Region Idea*, ed. Arif Dirlik (Boulder, CO: Westview Press, 1993), 251–82; "Mexico," in *The Encyclopedia of the Chinese Overseas*, ed. Lynn Pann (Cambridge, MA: Harvard University Press, 1998), 256–58; "On Coolies and Shopkeepers: The Chinese as Huagong (Laborers) and Huashang (Merchants) in Latin America/Caribbean," in *Displacements and Diasporas: Asians in the Americas*, ed. Wanni W. Anderson and Robert G. Lee (New Brunswick, NJ: Rutgers University Press, 2005): 78–111; "Multiculturalism in Latin American Studies: Locating the 'Asian' Immigrant; or, Where Are the Chinos and Turcos?," *Latin American Research Review* 44, no. 2 (2009): 235–42; "Latin America in Asia-Pacific Perspective," in *Asian Diasporas: New Formations, New Conceptions*, ed. Rhacel Parreñas and Lok Siu (Stanford, CA: Stanford University Press, 2007), 29–62; "Ceremonia Solemne de Recepción como Académica Corresponsal en Estados Unidos: Evelyn Hu-DeHart: Petición de perdón en Torreón y memoria histórica de los chinos en México," filmed January 2023 at Academia Mexicana de la Historia, Mexico City, https://www.youtube.com/watch?v=Qk46Uk0fkas.
18. Daniel Masterson with Sayaka Funada-Classen, *The Japanese in Latin America* (Champaign: University of Illinois Press, 2004); Robert Chao Romero, *The Chinese in Mexico, 1882-1940* (Tucson: University of Arizona Press, 2010); Grace Peña Delgado, *Making the Chinese Mexican: Global Migration, Localism, and Exclusion in the U.S.–Mexico Borderlands* (Stanford, CA: Stanford University Press, 2012); Jeffrey Lesser, *A Discontented Diaspora: Japanese Brazilians and the Meanings of Ethnic Militancy, 1960-1980* (Durham, NC: Duke University Press, 2007) and his edited book *Searching for Home Abroad: Japanese Brazilians and Transnationalism* (Durham, NC: Duke University Press, 2003); Julia María Schiavone Camacho, *Chinese Mexicans: Transpacific Migration and the Search for a Homeland, 1910-1960* (Chapel Hill: University of North Carolina Press, 2012); Elliott Young, *Alien Nation: Chinese Migration in the Americas from the Coolie Era through World War II* (Chapel Hill: University of North Carolina Press, 2014); Erika Lee, *The Making of Asian America: A History* (New York: Simon & Schuster, 2015); Fredy González, *Paisanos chinos: Transpacific Politics among Chinese Immigrants in Mexico* (Oakland: University of California Press, 2017).

further their interests; how they formed hybrid societies wherever they settled; or how the receiving societies excluded or integrated them, even though all this has been taken into consideration. Instead, it takes inspiration from the way in which Henry Yu and Elizabeth Sinn set out the approach to be adopted in the Crossing Seas series, of which this book forms a part, privileging "the journeys themselves, in multiple directions and at different times, between multiple locations, and what these journeys meant to the migrants as they crossed the seas."[19] Rather than centering solely on Chinese migrants and networks, this book concentrates on the maritime networks and the vessels that transported diverse people across the seas in order to show how the interests of travelers from one region sometimes coincided with those of other nationalities, ethnic backgrounds, and socioeconomic classes and how, at times, they collided, not only with other ethnicities but also with members of the same ethnic group or diaspora. By not concentrating solely on Chinese, as diaspora studies tend to do, but rather on steamship travelers, sailors, diplomats, port officers, and other professions that made travel possible, it becomes evident that the coincidence or divergence of interests is based not only on nationality or "race" but also on class, profession, and personal or group goals at specific periods of time. Since those who traveled across the ocean were overwhelmingly male and no records of female transoceanic travelers to and from Mexico for the studied period were found, gendered differences are not part of this study but remain an important pending subject.[20]

The Importance of Pacific Steamships in the Nineteenth Century

In its beginnings as a field, global studies concentrated on the Atlantic and the rise of British and US imperial interests, routes, and exchanges. Dirk Hoerder criticized this Atlanto-centric approach and advocated for an actual global vision that did not preconceive of the Atlantic as the avant-garde of what happened in the rest of the world. With this in mind, he identified five periods of global migration, the third of which frames the exchanges discussed in this book: in the nineteenth century, intercontinental migration systems formed when people moved in response to the demands for labor and the need for jobs created in the context of industrialization.

---

19. Henry Yu and Elizabeth Sinn, "Foreword," in *Returning Home with Glory: Chinese Villagers around the Pacific, 1849 to 1949*, by Michael Williams (Hong Kong: Hong Kong University Press, 2018), vii.
20. For a fascinating first-person account of the transpacific travels of a First Nation woman, see Margarita James, "My Transpacific Life," *BC Studies* 204 (Winter 2020): 139–50. For an excellent study on female transpacific travelers for an earlier period than the one covered in this book, see chap. 1 on Catarina de San Juan's journey in the seventeenth century in Seijas, *Asian Slaves*, 8–31. For a later period, see parts III and IV on the Mexican women married to Chinese men who were expelled from the country in the 1930s and who ended up living in Guangdong in Schiavone, *Chinese Mexicans*, 103–73. For the transatlantic world, see Adriana Méndez Rodenas, *Transatlantic Travels in Nineteenth-Century Latin America: European Women Pilgrims* (Lanham, MD: Bucknell University Press/Rowman & Littlefield, 2014) and Sara Beatriz Guardia, ed., *Viajeras entre dos mundos*, (Dourados: Ed. UFGD, 2012). The latter also includes an account of a trip along the Pacific coast of Mexico.

At the heart of transpacific movement between the 1830s and the 1920s, according to Hoerder, was the Asian contract labor system, because once slavery was abolished, the imperial powers found in Asian indentured servitude the cheap labor needed to industrialize their Pacific colonial domains.[21] Adam McKeown refined this argument by suggesting that a global perspective "provides insight not only into the global reaches of an expanding industrial economy, but also into how this integrative economy grew concurrently with political and cultural forces that favored fragmentation into nations, races, and perceptions of distinct cultural regions."[22] He coincided with Hoerder in seeing industrial transformations at the heart of human migrations in the nineteenth century, but he criticized him for considering mostly the industrial needs of European empires. McKeown instead argued that industrialization in Asia (and I would add Latin America as well) beyond European concerns also explained human movements because, after all, "non-Europeans were very much involved in the expansion and integration of the world economy."[23] As Mariano A. Bonialian pointed out when studying the centrality of Hispanic America in global exchanges between 1580 and 1840, "Hispanic America has exercised its own agency, one that isn't recognized in global history."[24] This book adds to these discussions by including the diverse industrial needs and interests of Mexicans as factors that encouraged transpacific mobility, while keeping in mind the centrality of Asian migrants suggested by Hoerder and McKeown, in particular those from the province of Guangdong.

In this regard, the works of Henry Yu and Elisabeth Sinn have been of particular use as they have explained both the rise of Hong Kong as a transoceanic hub and the centrality of Cantonese in transpacific exchanges as of the second half of the nineteenth century. As Yu pointed out, hundreds of thousands of migrants had departed from the southern ports of Guangdong and Fujian provinces since the fifteenth century for a diverse array of destinations. But it was after the takeover of Hong Kong by the British in 1842, as part of the settlement with the Qing Empire over the opium trade, that this port monopolized transpacific travel, so that those from nearby Cantonese-speaking counties, with their aspirations for a better life, formed the large majority of travelers crossing the ocean. Together, they created "a coherent century-long migration process that was persistent, recurring, and unique in its effect on global history."[25] On her part, Elisabeth Sinn explained how Hong Kong transformed from a small British port centered on the transportation of raw opium

---

21. See chap. 15 in Dirk Hoerder, *Cultures in Contact: World Migrations in the Second Millennium* (Durham, NC: Duke University Press, 2002), 366–404.
22. Adam McKeown, "Global Migration, 1846–1940," *Journal of World History* 15, no. 2 (2004): 156.
23. McKeown, "Global Migration," 171.
24. Mariano A. Bonialian, *La América española: entre el Pacífico y el Atlántico: globalización mercantil y economía política, 1580-1940* (Mexico City: El Colegio de México, 2019), 16.
25. Henry Yu, "Unbound Space: Migration, Aspiration, and the Making of Time in the Cantonese Pacific," in *Pacific Futures: Past and Present*, ed. Warwick Anderson, Miranda Johnson, and Barbara Brookes (Honolulu: University of Hawai'i Press, 2018), 178.

to westward cities to a global shipping hub thanks to the social, economic, and cultural exchanges promoted by hundreds of thousands of Cantonese free migrants who traveled through Hong Kong to various ports, particularly San Francisco, after the Gold Rush.[26] Her work was a necessary complement to the numerous studies focused on Cantonese indentured experiences. This book explores the journeys of some of those Cantonese who traveled to Mexican ports as part of the process elucidated by Yu and Sinn.[27]

Sinn has also provided a key concept for this book, that of in-between places. Arguing that "migration is seldom a simple, direct process of moving from Place A to Place B," she finds instead a process of repeated, even continuous, movement, along with the appearance of hubs that "witness the coming and going of persons and things"; such hubs provide infrastructure to continue traveling as well as to maintain ties with the homes left behind. In order to "accentuate the sense of mobility," Sinn prefers to call Hong Kong an "in-between place" and acknowledges that other locations may also be described this way.[28] Mexican historian Karina Busto Ibarra also defined ports as dynamic sites where regional economies were articulated and inserted into global processes.[29] With these authors in mind, I argue that by conceiving of not only the ports that the transpacific steamships described in this book regularly visited, such as Hong Kong, Yokohama, San Francisco, Manzanillo, and Salina Cruz, but also the steamships themselves, as in-between places, we are better able to grasp the sense of mobility that characterized the Pacific basin in the nineteenth century. After all, the steamships as well as these ports were the first places of exposure and contact between what had been left behind and what was to come. In effect, during the few weeks that the transpacific journeys lasted, travelers were exposed to ideas, hierarchies, experiences, and relations that had to do with

---

26. See Elisabeth Sinn, *Pacific Crossing: California Gold, Chinese Migration, and the Making of Hong Kong* (Hong Kong: Hong Kong University Press, 2013).
27. The Migration Law of 1926 required foreigners residing in Mexico to register their personal information with the federal government in the recently created National Registry for Foreigners or *Registro Nacional de Extranjeros* (RNE). The *Archivo General de la Nación* safeguards some 14,000 RNE cards of Chinese residents who entered the country between 1895 and 1949. I reviewed some 3,000 from those who landed in the port of Manzanillo. While the large majority simply cited China as their place of origin, those that included a city cited Kaiping (Hoiping), Toisan (Taishan), and Hong Kong, confirming that Mexico was part of the "Cantonese Pacific" described by Henry Yu. Ruth Mandujano López, "La migración interminable, cantoneses en Manzanillo," *Legajos. Boletín del Archivo General de la Nación* 1 (July–September 2009): 48–49.
For a general overview of the information found in the over 14,000 Chinese records, see Roberto Ham Chande, "La migración china hacia México a través del Registro Nacional de Extranjeros," in *Destino México: Un estudio de las migraciones asiáticas a México, siglos XIX y XX*, ed. María Elena Ota Mishima (Mexico City: El Colegio de México, 1997), 167–88.
28. Sinn, *Pacific Crossing*, 9. I have discussed Sinn's notion of the in-between in Mandujano, "Migración interminable," 48, 58 as well as in "Cantoneses en Manzanillo: la importancia del "lugar de en medio" en el proceso migratorio," in *Tierra receptora y espacios de apropiación. Extranjeros en la historia de México, siglos XIX y XX*, ed. Martín López and Marcela Martínez (Zamora: El Colegio de Michoacán y El Colegio de San Luis, 2015), 321–36.
29. Karina Busto Ibarra, *El Pacífico mexicano y sus transformaciones: integración marítima y terrestre en la configuración de un espacio internacional, 1848–1927* (Mexico City: El Colegio de México, 2022), 19.

the places they left behind as well as the places they were going to. As the chapters in this book show, steamships were in-between places where people's experiences both shaped and were shaped by power relations as defined by imperial, national, and capitalist enterprises as well as by their own family and interpersonal interests and negotiations. In addition, understanding the voyages themselves can also shed light on what happened to people as they landed (or were refused landing), settled, and/or continued to move around.

Travel and travel writing studies also offer conceptual support for the book as they remind readers that movement is an essential trait in human history. "Whether we travel to foreign lands or just across the room, we all journey and from our journeying define ourselves."[30] As movement defines our lives, our expressions are full of terms and metaphors related to it, for instance, "as your eyes travel through this introduction, as you read my words going across the page, keep track of how often you get up from this reading to move about the room and refresh yourself from your otherwise stationary task."[31] This forces us to rethink the discursive rootedness and politics of belonging that national histories tend to enforce. Another important premise to highlight from this field is that travel and descriptions of travel have been at the core of the inequalities related to nation and empire-making and the reproduction and expansion of capitalism, so there is an inextricable link between mobility and power that is in turn permeated by class, ethnicity, profession, gender, age, place, identity, scientific paradigms, and other distinctions.[32] It is thus important to pay attention not only to who travels and under what conditions but also to who speaks about the trip when travel narratives are the main source. Since many of the primary accounts from this book come from national archives or from travelers whose mobility was sponsored either by merchant maritime companies or national and imperial governments, the asymmetries that they directly or indirectly promoted should not be overlooked.

Jerry H. Bentley's notion of a maritime region, Steven Vertovec's definition of social networks, and Bruno Latour's actor network theory are also central concepts in this book. Bentley's work has been the pillar for the idea of a Pacific region for numerous historians. According to him, sea and ocean basins become useful categories once "human societies engage in interactions across bodies of water and they become a less useful focus as societies pursue their interests through other spaces."[33] For Bentley, integration, defined as a "historical process that unfolds when

---

30. Susan L. Roberson, "Defining Travel: An Introduction," in *Defining Travel: Diverse Visions*, ed. Susan L. Roberson (Jackson: University Press of Mississippi, 2007), xi.
31. Roberson, "Defining Travel," xi.
32. See the texts by different authors that compose Roberson, *Defining Travel* and those from Julia Kuehn and Paul Smethurst, eds., *Travel Writing, Form, and Empire* (New York: Routledge, 2009). See also Mary Louise Pratt, *Imperial Eyes: Travel Writing and Transculturation* (New York: Routledge, 1992).
33. Jerry H. Bentley, "Sea and Ocean Basins as Frameworks of Historical Analysis," *Geographical Review* 89, no. 2 (1999): 217. Bentley recognizes Fernand Braudel's *La Méditerranée et le monde méditerranéen à l'époque de Philippe II* as the foundational study that worked with a maritime region.

cross-cultural interactions bring about a division of labor between and among interacting societies or when they facilitate commercial, biological, or cultural exchanges between and among interacting societies on a regular and systematic basis,"[34] is a necessary element for a maritime region to exist. Bentley admits that there are never absolute and fixed boundaries. Rather, it is the particular circumstances of each case that determine the limits of a maritime region. This study shows that through the efforts and travels of various Mexicans, Japanese, and Cantonese, the Mexican ports of Manzanillo and Salina Cruz became integrated into a transpacific maritime region that they helped to shape. Vertovec, for his part, talks about space in terms of social networks, that is, each person is seen as a node linked with others to form a network.[35] Latour expanded this concept to include nonhuman and non-individual actors—or actants—because they also participate in the functioning of a certain social order.[36] The transpacific network that this work studies is therefore formed by people—travelers, diplomats, crews, and others who made transpacific travel possible; ports—notably Hong Kong, Yokohama, San Francisco, Manzanillo, Salina Cruz, and Panama; and technological devices—the most important of which is the steamer.

Tatiana Seijas stresses another reason why a Pacific world vision that takes into consideration what happens at sea is of worth. She states that such a paradigm "offers a path for studying Asian migrations as the multi-directional flow that it was . . . [and] encourages historians to make non-territorial connections and showcase how water currents shaped peoples and environments in all cardinal directions."[37] In this regard, the research on the Pacific Mail Steamship and other companies that operated from San Francisco by John Haskell Kemble, William Kooiman, and particularly E. Mowbray Tate also serves as a key point of departure for this study.[38] By tracing the history of specific companies and their steamers, these historians succeeded in showing the dynamism and mobility that has characterized the Pacific and pointed to the centrality of vessels in the construction of a maritime region. In regard to Mexican mercantile maritime relations, this book benefited from the early

---

34. Bentley, "Sea and Ocean Basins," 218.
35. Networks are characterized by size, defined by the number of participants; density, or the extent to which every one of the nodes contacts the others; multiplexity, or the degree to which relations between participants include overlapping institutional spheres; clusters or cliques, or the specific area of a wider network with higher density than that of the network as a whole; durability or length; and frequency, or the regularity of contacts within the network. See Steven Vertovec, "Migration and Other Modes of Transnationalism: Towards Conceptual Cross-Fertilization," *International Migration Review* 37, no. 3 (Fall 2003): 646–47.
36. Bruno Latour, *Reassembling the Social: An Introduction to Actor-Network Theory* (Oxford: Oxford University Press, 2005).
37. Seijas, "Asian Migrations," 574.
38. John Haskell Kemble, "The Genesis of the Pacific Mail Steamship Company," *California Historical Society Quarterly* 13, no. 4 (December 1934): 386–406 and "Pacific Mail Service between Panama and San Francisco, 1849–1851," *Pacific Historical Review* 2, no. 4 (December 1933): 405–17 (among others); William Kooiman, "Grace's Pacific Mail, 1915–1925," *Journal of the Puget Sound Maritime Historical Society* 21, no. 1 (September 1987): 3–20; E. Mowbray Tate, *Transpacific Steam: The Story of Steam Navigation from the Pacific Coast of North America to the Far East and the Antipodes, 1867–1941* (New York: Cornwall Books, 1986).

surveys of Juan de Dios Bonilla, Enrique Cárdenas de la Peña, Inés Herrera, Vera Valdés Lakowsky, and especially the more recent work of Karina Busto Ibarra. Busto Ibarra carefully studied how Mexican ports became incorporated after the Gold Rush into what she has described as the San Francisco–Panama axis, part of a larger set of transformations experienced throughout the Pacific due to the introduction of steam technologies.[39] This book adds to this body of work by determining the routes and studying the companies, steamships, and people that circulated between Mexican ports and Hong Kong and by examining some of the challenges, experiences, and consequences of this constant movement of people and vessels across the ocean; its focus is on personal lives as well as national and imperial imaginings and policies.

An actual maritime journey on the Pacific gave me the last intellectual push I needed for conceiving this book. Some time ago, I embarked on a memorable boat ride to Yuquot, where local leader Margarita James explained how the Mowachaht/Muchalaht First Nations (MMFN) have been part of global history and how their territory became a transpacific hub in the eighteenth century.[40] Her views on the subject reflect what one of the welcoming plaques explain to visitors: "Explorers and traders were attracted to this safe harbour, which they called Friendly Cove. As a result, Yuquot, also known as Nootka, developed into an important center of trade and diplomacy, and it was briefly the site of Spain's only military establishment in present-day Canada. Yuquot became the focal point of the Nootka Sound Controversy of 1789–1794, when the rival interests of Great Britain and Spain brought those countries to the brink of war."[41] After his own visit to Yuquot, historian John Price wrote an article that explores the transpacific mobility of the MMFN and of Chinese, whom he described as "essential to the operations of many vessels plying the Pacific." He concluded that there have been "multiple, overlapping networks of migration and trade in the Pacific" and that "the overlapping tides of Indigeneity, imperialism, and migration/diaspora should be given their proper weight."[42] Just like Yuquot, the Mexican coast found itself at the heart of the rival imperial interests of Spain and Great Britain. Chinese also played a key role in

---

39. Juan de Dios Bonilla, *Historia marítima de México* (Mexico City: Litorales, 1962); Vera Valdés Lakowsky, "Cambios en las relaciones transpacíficas: del *Hispanis Mare Pacificum* al Océano Pacífico como vía de comunicación internacional," *Revista de Estudios de Asia y África* 53, no. 1 (1985): 58–81; Enrique Cárdenas de la Peña, *Historia de las Comunicaciones y Transportes en México: Marina mercante* (Mexico City: Secretaría de Comunicaciones y Transportes, 1988); Inés Herrera, "Comercio y comerciantes de la costa del Pacífico mexicano a mediados del siglo XIX," *Historias* 20 (September 1988): 129–36, https://revistas.inah.gob.mx/index.php/historias/article/view/14909; Karina Busto Ibarra, "El espacio del Pacífico mexicano: puertos, rutas, navegación y redes comerciales, 1848–1927" (PhD diss., El Colegio de México, 2008); Busto, *Pacífico mexicano*.
40. Margarita James is the president of the Land of Maquinna Cultural Society, a nonprofit arm of the Mowachaht/Muchalaht First Nations (MMFN) whose purpose is to promote Yuquot's heritage.
41. Copied by the author from the original plaque and also cited in John Price, "Relocating Yuquot: The Indigenous Pacific and Transpacific Migrations," *BC Studies* 204 (Winter 2019–2020): 22–23.
42. Price, "Relocating Yuquot," 41–42.

linking it with the transpacific world. This book explains how parts of the Mexican coastline reconnected with Asia during the transition from Spanish to Anglo imperialism, from colony to nation, and from sail to steam in the nineteenth and early twentieth centuries. Recognizing the voices and travels of First Nations inhabiting the Mexican coastline remains one of the pending tides to give voice to and to analyze.

Chapters, Methodology, and Sources

Methodologically, this book begins from a basic premise of transnationalism that maintains that researchers have to be as mobile as the subjects and flows they study. As Tatiana Seijas argues, "scholars must engage varied historiographies, gather documentation in distant archives, and read different languages."[43] In this sense, this research has sought to include archival sources from almost as many places as are mentioned in the book. As a consequence, the materials used come from over twenty personal, municipal, provincial, and federal archives and libraries in the following locations: Hong Kong, San Francisco, Mazatlan, Manzanillo, Colima, Mexico City, Oaxaca, Seville, and London. Archival sources range from the 1820s to the 1920s and include personal memoirs and letters, diplomatic correspondence, maps, newspapers and other periodicals, journals, censuses, books, and brochures, mostly in English and Spanish. As such, it incorporates the voices of Mexicans, Cantonese, Japanese, British, and bureaucrats from the United States that spoke or were translated into those languages.

The book is divided into six chapters, each centering on a specific steamship that encapsulates the state of transpacific journeys "in-between" Mexican ports, Yokohama, and/or Hong Kong during the specific decade analyzed in the chapter. It is therefore not an exhaustive review of transpacific travels in the age of steam, but rather a selective study of key steamship journeys and companies between the 1860s and the 1910s.[44] The time frame corresponds to the beginning of regular transpacific passenger services by steamship and the pause of the services that happened in the context of the First World War. In order to favor a more literary flow that is accessible to everyone and so that interested readers have additional bibliography for the multiple topics that are not at the center of the book but intersect with its themes, academic references have been, for the most part, taken out of the main text and put into the footnotes.

Chapter 1 takes as its point of departure Tatiana Seijas's argument for "a Pacific World perspective in the longue-durée" when analyzing the history of the people who crossed the ocean because "studying their experiences from a frame of four

---

43. Seijas, "Asian Migrations," 577.
44. Even though the flow of commodities, notably of silver, is sometimes mentioned, it is not part of this study because it requires its own contextualization and research that goes beyond the scope of this book.

hundred years (1500–1800s) allows for more nuanced understandings of the hemispheric and transpacific connectivities of this human story before the reification of historical narratives centered on the nation state."[45] With this in mind, the chapter starts with the Manila–Acapulco route inaugurated by the galleon *San Pedro* in 1565, which connected the two continents on their northern hemispheres on a regular basis for the first time in history, between the sixteenth and the early nineteenth centuries. It then moves to examine the transition from sail- to steam-powered navigation in the Pacific, which came hand in hand with the transition from Spanish to Anglo imperialism and from colonial to national status. It explores how all these transformations affected the Mexican coastline and its transpacific connections. It ends with the trip of *Colorado* in 1867, which inaugurated regular steamship passenger services across the north Pacific.

Chapter 2 centers on the travels aboard *Vasco de Gama* made in 1874 by the first official Mexican delegation to visit Asia. The impressions and the experiences of the scientists who comprised it enable a discussion of the intricacies of the trip during the first decade of regular transpacific passenger services, when steam technologies were not yet fully developed. The analysis of the diaries of the travelers shows how they conceived of the importance of transoceanic connections between Mexico, Japan, and China for the development of scientific and national objectives as defined by the national government with which they were associated. It also shows how power and mobility related to the interconnected notions of race, nation, science, and empire at the time.

Chapter 3 studies *Mount Lebanon* and the formation of the first Mexican steamship venture ever created to establish direct maritime routes between Mexico and Asia in 1884. Using mostly government records and letters found in British, Mexican, and Hong Kongese archives, this case study reveals the coincidence and divergence of interests between Mexican, Chinese, and British diplomats and businessmen as they all tried to partake in the lucrative business of transoceanic passenger transportation. It also shows that, even if journeys did not actually materialize, their mere planning and negotiation mattered for personal, national, and imperial objectives.

Chapter 4 explains the coinciding interests of Japanese and Mexican diplomats and travelers that led to the arrival of the first Japanese colonists in Latin America aboard *Gaelic* in 1897. Using newspapers from the time as well as various secondary sources that have studied the period, the chapter delves into the parallel modernization projects, journeys, and diplomatic exchanges that made it possible for Mexico and Japan to establish relations in 1888 as well as into the subsequent travels by Japanese officers and eventually those of the first colonists to show how nations and empires have been created through movement.

---

45. Seijas, "Asian Migrations," 573.

Chapter 5 centers on the first successful regular direct service between Mexican and Asian ports created by the CCSC during the first decade of the twentieth century. Using mostly official correspondence found in Mexican and Hong Kongese archives as well as newspapers from the time, it exposes the coincidences and divergences between British, Mexican, and Chinese authorities and businessmen in relation to transpacific passenger services and how travelers from different classes (first, second, and steerage) experienced their trip and negotiated their entrance or refusal to the country.

Finally, Chapter 6 explains how the maritime system that connected Mexican Pacific ports with Hong Kong began to decay in the 1910s in the context of a series of transformations which included the Mexican Revolution and the First World War, the opening of the Panama Canal, and the replacement of steam by oil as the vessels' main fuel.

I referred earlier to the importance that the concept of "in-between" places has in my work in highlighting the sense of mobility and the connections to what was left behind and what was to come for the people and ships that are at the center of this book. But it goes beyond this. As someone who travels and encounters different places and ideas and feels connected to different worlds on a regular basis, I define myself as also being "in-between" and therefore my work does not aim to be definitive. As Susan L. Roberson concludes: "Traveling across disciplines and individual experiences, we find that travel itself is an unsettled term, one whose definition depends on the particular 'politics of location' of the writer . . . what constitutes mobility . . . is not so easily defined. Mobility, too, is mobile."[46] I therefore encourage readers and researchers to engage with and reassess my ideas according to your own backgrounds as you move through the chapters.

All voyages, all stories of voyages, have a beginning. Sometimes, for dramatic effect, stories of voyages begin at the end. Mine begins in-between.

---

46. Roberson, "Defining Travel," xxiii.

1
San Pedro, 1565 / Colorado, 1867
*From Sail to Steam**

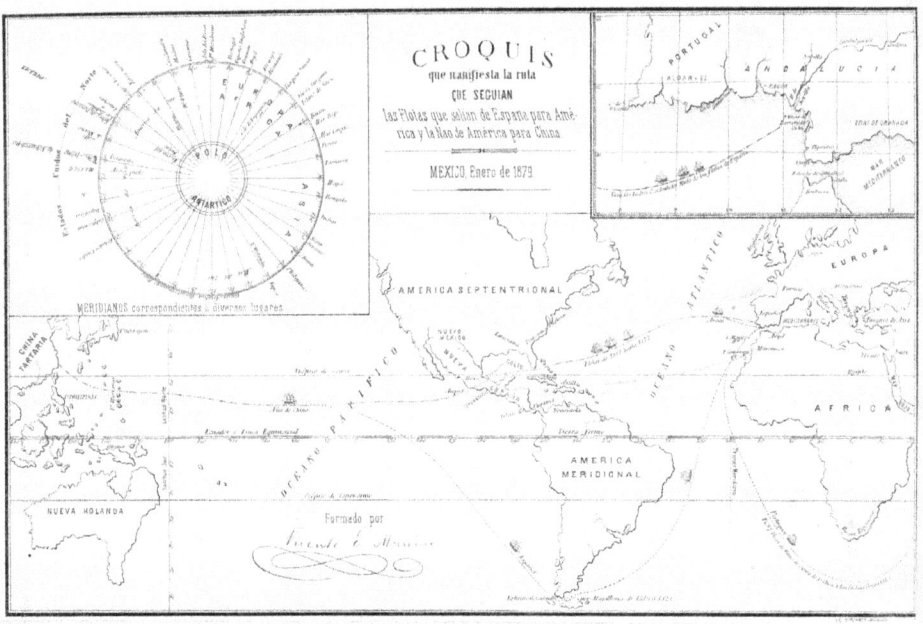

Map 1
Vicente E. Merino, *Ruta de las flotas de España para América y la Nao de América para China* [Route followed by the Spanish fleets to America and from America's Nao to China]. Map. Mexico City: H. Iriarte, 1879. From Mapoteca "Manuel Orozco y Berra," Internacionales 3. Accessed December 15, 2023. https://mapoteca.siap.gob.mx/coyb-int-m50-v3-0075/.

---

* Data from this and Chapter 3 appeared in the journal *Retos Internacionales*. I thank founder and director Gabriel Morelos Borja for permission to reuse it.

By the spring of 1560, Friar Andrés de Urdaneta had settled into his new life behind the thick walls of a peaceful Mexico City monastery. A latecomer to the cloth, he had joined only some seven years earlier at the ripe age of forty-five. His new life in relative obscurity as far away from a port as possible was due to choice more than chance. He bore the scars of a complex personal history: as a teenager he had joined José García Jofre de Loaysa's 1525 colonizing expedition to the Molucca islands, the world's fountainhead of precious clove, mace, and nutmeg. But since the Portuguese crown controlled the well-known African coastal route, the Spanish monarch sent Loaysa west across the Atlantic and Pacific oceans via the Strait of Magellan. The human cost was disastrous: only a handful of the 450 sailors survived. Urdaneta was shipwrecked and spent the better part of a decade fighting for his life in the Molucca amid shifting alliances among warring groups of locals, Muslims, Spanish, and Portuguese. He finally escaped and returned to Spain. He then recrossed the Atlantic and soon thereafter assimilated into the Order of Saint Augustine.[1]

Then the king's letter arrived.

In 1559, after several expeditions had failed to find a return route from Asia to the Americas,[2] King Phillip II—after whom the Philippines received its name—ordered New Spain's viceroy to dispatch a new expedition.[3] The viceroy, in turn, had the 500-ton *San Pedro* and the 400-ton *San Pablo* built on the shores of Navidad using local woods and additional materials brought all the way from Europe and across the narrow stretch of New Spanish land between Guaçaqualco in the Atlantic and Teguantepeque in the Pacific.[4] Despite Urdaneta's newfound vocation, the mission needed him; he was one of the few people alive who had circumnavigated the globe.

In the letter, the king requested that "you go on these vessels and follow the viceroy's commands," because, "according to the amount of information that you have said that you have about the ways of that land, and since you know how to navigate there, and being a good cosmographer, it would be of great effect that you were on those boats and that you participate in the navigation there, as part of your

---

1. The original writings by Urdaneta that describe the Loaysa expedition and his time in the Molucca islands are available at www.andresurdaneta.org, a webpage created by the government of Ordizia, Urdaneta's hometown, to commemorate the 500th anniversary of his birth; Luis Laorden Jiménez, *Navegantes españoles en el Océano Pacífico* (Madrid: Taograf, 2013), 52, 94–101; José Ramón de Miguel Bosch, "Andrés de Urdaneta y el tornaviaje," Urdaneta500, accessed November 21, 2023, http://www.andresurdaneta.org/urdaneta500/de/biografia.asp?cod=1754&nombre=1754&nodo=&orden=True&sesion=1.
2. Prior to 1565 Europeans knew how to sail from the American continent to Asia, but the route eastward had not been found as the wind and currents had only caused those who had tried it to shipwreck.
3. The Spanish originally divided their American colonies into two viceroyalties, New Spain in the north, with Mexico City as its capital, and Peru in the south, with Lima as capital.
4. With time, the names of these two places evolved to Coatzacalcos and Tehuantepec, two key ports in relation to transoceanic routes that will be discussed in Chapter 5.

service to God and us."⁵ Urdaneta had previously refused a similar proposal by the viceroy. This time he responded, "Given that my age is over fifty-two years, and my poor state of present health, and the many works I have passed since my youth, I was in need of spending the little time that remains in my life in peace and quiet. But considering the great zeal that your Majesty has for everything related to the provision of our Lord and augmented by your Catholic faith, I will make myself available for the work of this project."⁶

Miguel López de Legazpi's four-vessel fleet departed in November 1564 and arrived in Cebu, in the Philippines, in April 1565. Urdaneta was head cosmographer and one of a handful of Augustinian friars on board. Legazpi stayed, becoming the Philippines' first Spanish governor. He left the return trip in the hands of his eighteen-year-old grandson, Felipe de Salcedo, and Urdaneta. Together with some 200 other men, they departed on June 1 aboard *San Pedro*, taking some 130 days to reach the Spanish American coast. Only eighteen healthy sailors were standing on arrival, but they landed with clear coordinates of the Philippines to Acapulco return route.⁷ *San Pedro* thus inaugurated one of the longest-lasting trade passages in history, mainly characterized by the transportation of Chinese manufactured products in exchange for Mexican silver coins aboard galleons commanded by Spanish officers but manned mostly by Chinese and other Asian populations. This flow of people and merchandise has often been described as the birth of globalization.⁸ The circuit lasted as long as the Spanish controlled their American colonies. In the context of the Mexican War of Independence, the last of the galleons sailed in 1815, leaving the continent's relationship dependent on irregular transpacific crossings as well as those via the Atlantic and Indian oceans.

---

5. "1559 Carta del Rey a Fr. Andrés de Urdaneta, encargándole que se embarque en los navíos que el Virrey de Nueva España enviará al descubrimiento de las islas Filipinas," Memoria Política de México, accessed November 21, 2023, https://www.memoriapoliticademexico.org/Textos/1Independencia/1559-C-FII-AU-R.html. Translations from Spanish documents into English by Martin McLennan and the author.
6. Archivo General de Indias (AGI), Fondo México, 19, N. 23, Carta del virrey Luis de Velasco y Ruiz de Alarcón, anexo "Fray Andrés de Urdaneta a S.M., sobre su ida a la jornada que se ha de hacer a las islas del Poniente. México, 28-V-1560."
7. Laorden, *Navegantes españoles*, 95–101.
8. Evelyn Hu-DeHart, "Ceremonia Solemne de Recepción como Académica Corresponsal en Estados Unidos: Evelyn Hu-DeHart: Petición de perdón en Torreón y memoria histórica de los chinos en México," filmed January 2023 at Academia Mexicana de la Historia, Mexico City, https://www.youtube.com/watch?v=Qk46Uk0fkas; Peter Gordon and Juan José Morales, *La plata y el pacífico. China, Hispanoamérica y el nacimiento de la globalización, 1565–1815* (Madrid: Siruela, 2022), 22–23; Déborah Oropeza, *La migración asiática en el virreinato de la Nueva España: un proceso de globalización, 1565–1700* (Mexico City: El Colegio de México, 2020), 20; Mariano A. Bonialian, *La América española: entre el Pacífico y el Atlántico: globalización mercantil y economía política, 1580–1840* (Mexico City: El Colegio de México, 2019), 14; Dennis Flynn and Arturo Giráldez, "Los orígenes de la globalización en el siglo XVI," in *Oro y plata en los inicios de la economía global: de las minas a la moneda*, ed. Bernd Hausberger and Antonio Ibarra (Mexico City: El Colegio de México, 2014), 29–76; Dennis Flynn and Arturo Giráldez, "Cycles of Silver: Globalization as Historical Process," *World Economics* 3, no. 2 (April–June 2002): 1–16. Javier Mejia offers a different perspective, arguing that "the relevance of the Manila Galleon should not be understood as the starting point of globalization. Instead, the Manila Galleon should be seen as a fundamental step in a gradual process of global integration." Javier Mejia, "The Economics of the Manila Galleon," *Journal of Chinese Economic and Foreign Trade Studies* 15, no. 1 (2021): 56.

Regular north Pacific routes did not restart until 1867, this time aboard steamships. *Colorado*, a purpose-built 3,728-ton, 340-foot-long side-wheeler, was the first. It departed San Francisco on January 1, with only fifty-one first-class passengers in its fifty-three staterooms. Similarly, below deck there were only 200 Chinese passengers in a steerage that accommodated five times as many; 1,000 barrels of flour hardly filled its hold. These numbers belied the magnitude of the sailing, as the three correspondents for the nation's most important newspapers and the presence of the president of the New York Chamber of Commerce foretold its legacy.[9] At noon, *Colorado* blew its whistles and the gathered crowd burst into shouts of euphoria. As four cannon salutes responded to its call, *Colorado* set course for Yokohama and Hong Kong, a giant plume of black exhaust belching out of its smokestack. Thus departed the inaugural Pacific Mail Steamship Company (PMSS) vessel—and the first-ever regular steamship and passenger liner—to cross the northern axis of the world's largest ocean.[10]

This chapter explains the long history of regular transpacific exchanges between eastern Asia and Mexico from the sailing of *San Pedro* in 1565 to the voyage of *Colorado* in 1867. It delves into the transition from sail to steam navigation and highlights how this affected coastal exchanges, with Manila and Acapulco giving way to Hong Kong, Yokohama, and San Francisco as the primary transpacific nodes in the emerging context of Anglo-American imperialism. With the founding of PMSS and particularly after the company opened its transpacific line in 1867, San Francisco became the intermediary between Mexico and Asia.[11] Finally, the chapter also exposes the important role played by Chinese throughout the centuries, first as suppliers and consumers of the bulk of merchandise and coins traveling in the galleons and then as steerage passengers, the main business of the transpacific steamships that followed *Colorado*.

Pre-Steam Transpacific Contacts

Despite the Pacific Ocean's overwhelming dimensions—its size is equivalent to a third of the Earth's surface—the peoples living on its opposing coasts have bridged the space separating them for centuries, particularly in the southern hemisphere.[12]

---

9. "New York to China," *The Daily Bee*, January 2, 1867, 2; A. H. Cathcart, "Pacific Mail—Under the American Flag Around the World," *Pacific Marine Review*, July 1920, 56; TheShipsList, "Colorado."
10. E. Mowbray Tate, *Transpacific Steam: The Story of Steam Navigation from the Pacific Coast of North America to the Far East and the Antipodes, 1867–1941* (New York: Cornwall Books, 1986), 23–25.
11. Karina Busto Ibarra, *El Pacífico mexicano y sus transformaciones: integración marítima y terrestre en la configuración de un espacio internacional, 1848–1927* (Mexico City: El Colegio de México, 2022), 190.
12. See Ben Finney, "The Other One-Third of the Globe," *Journal of World History* 5, no. 2 (1994): 273–97; P. S. Bellwood, "The Peopling of the Pacific," *Scientific American* 243, no. 5 (1980): 1–17; Stephen Jett, *Crossing Ancient Oceans: Voyages to the Americas before Columbus* (New York: Copernicus, 2007); and part one of William F. McNeil, *Visitors to Ancient America: The Evidence for European and Asian Presence in America Prior to Columbus* (Jefferson, NC: McFarland & Company, 2005). George F. Carter has an interesting essay that focuses on the transfer of plants among the southern Pacific peoples to demonstrate that they were in

Nevertheless, systematic exchanges between what we know today as Mexico and the Asian continent only began in the sixteenth century in the context of fierce competition for the control of the Asian spice trade. With the traditional land routes under Muslim control, Europeans, particularly the recently unified Christian kingdoms of Portugal and Spain, were eager to find a way around them by sea. The Portuguese looked south and soon dominated the route bordering Africa and into the Indian Ocean. The Spanish kings had no choice but to search for an unknown, westward transatlantic route. The resulting 1492 trip by Christopher Columbus accidentally exposed an entire new continent that they soon began to explore, conquer, and colonize.[13]

It took the Spanish a few decades to find their way across the Americas into the Pacific and Asia.[14] Their first state-sponsored expedition departed from Seville in August 1519. It navigated the Atlantic and bordered the South American coast, then sailed across the Pacific to today's Philippines, where commander Ferdinand Magellan met his death. Sebastián Elcano replaced him and returned to Seville aboard the galleon *Victoria* with only eighteen of the initial 250 or so sailors, via the Indian Ocean, in September 1522.[15] Elcano was back at sea in a failed 1525 expedition commanded by friar García Jofre de Loaysa, where almost everyone, including Elcano and Loaysa, died. One of the few survivors, Andrés de Urdaneta, captained a more successful trip in 1565.[16] Up to that point, Europeans knew how to sail from the American to the Asian continent westward, but headwinds and currents impeded their return using the same route; the eastward leg had not yet been mapped. As mentioned earlier, Urdaneta became the first to do so in an expedition that departed on June 1 from the port of Cebú, south of Manila,[17] aboard *San Pedro*, described as follows by the chief pilot Esteban Rodríguez: "The captain's nao was ready to sail, well-stocked with more than eight months' worth of bread and rice and corn and broad beans and garbanzos and oil and vinegar and wine, and 200 barrels of water; there were two hundred people on board with ten soldiers and two priests."[18] *San Pedro* sailed northeast along the Asian coast and crossed the open sea

---

regular and systematic contact before European arrival; see "Movement of People and Ideas across the Pacific," in *Plants and the Migrations of Pacific Peoples: A Symposium*, ed. Jacques Barrau (Honolulu: Bishop Museum Press, 1961), 7–22.

13. Gordon and Morales, *La plata y el Pacífico*, 10–11, 25–27; Erika Lee, *The Making of Asian America: A History* (New York: Simon & Schuster, 2015), 17–18.
14. The original name given by the Spanish to the Pacific was *Mar del sur* or Southern Sea.
15. Juan Parejo, "Una versión inédita de la vuelta al mundo de Magallanes y Elcano," *Diario de Sevilla*, November 7, 2019, https://www.diariodesevilla.es/sevilla/version-inedita-vuelta-mundo-Magallanes-Elcano_0_14074 59578.html; Laorden, *Navegantes españoles*, 46–48. Paradoxically, it was a Portuguese sailor, Magellan, who commanded the voyage for the rival crown of Spain. Gordon and Morales, *La plata y el Pacífico*, 28.
16. Laorden, *Navegantes españoles*, 52.
17. Gordon and Morales, *La plata y el Pacífico*, 28–37.
18. Original in Spanish cited in Blas Sierra de la Calle, "La expedición de Legazpi-Urdaneta (1564–1565). El tornaviaje y sus frutos," in *Instituto de Historia y Cultura Naval. Cuaderno monográfico n° 58* (Madrid: Ministerio de Defensa, 2009), 147, https://armada.defensa.gob.es/archivo/mardigitalrevistas/cuadernosihcn/58cuaderno/cap05.pdf.

between parallels thirty-six and forty.[19] Some 130 days later, after several storms and mishaps, including Rodríguez's death, it reached the Spanish American coast.[20] One of the few remaining healthy sailors described the scene as follows:

> We awoke in the Port of Navidad and, at that moment, I looked at my map and saw that we had traveled some 1892 leagues (13,000 kilometers) from the port of Cebu to the port of Navidad, and at that moment I went to the captain and asked him where should I take the ship . . . and he ordered me to take it to the Port of Acapulco. I obeyed his order, although there were not more than eighteen men who could work the nao, because the rest were sick, and another sixteen had died. We arrived in the Port of Acapulco on the 8th of the current month of October due to the very hard work of all the people.[21]

The *San Pedro* sailing, despite its fatalities, inaugurated one of the most long-lasting trading routes in history, that of the so-called Manila galleons.

Two main factors that contributed to the success of this route were the consolidation of Spanish imperialism on both sides of the Pacific and Chinese interest in American silver.[22] In effect, by the second half of the sixteenth century, Spanish armies, Indigenous allies, and epidemic disease had combined to defeat two of the main Indigenous empires of the time: the Incas and the Aztecs. On the appropriated lands, the crown established two main administrative centers, the viceroyalties of Peru along most of South America's Pacific coast and New Spain, covering roughly current Mexico, most of Central America and the Antilles. The latter included the Philippines. The two American viceroyalties happened to possess the largest supplies of silver known at the time,[23] which coincided with the Ming dynasty's enforcement of a series of fiscal reforms that included the payment of taxes in silver.[24] China's reliance on silver was so significant that calculations suggest that at least a third of the silver mined in the Spanish Americas during colonial times ended up there.[25]

---

19. Andrés de Urdaneta found the route back by sailing northeast of the Philippines as far as parallel 40 following the "Kuroshio" or Black current, also called the Japan current, to California and then south along the coast to Acapulco. Rubén Carrillo Martín, "Los 'chinos' de Nueva España: Migración asiática en el México colonial," *Millars Espai i Història* 21, no. 39 (January 2015): 21, https://core.ac.uk/download/pdf/132357334.pdf.
20. Sierra, "La expedición," 147; Gordon and Morales, *La plata y el Pacífico*, 36.
21. Original in Spanish cited in Sierra, "La expedición," 150.
22. For the purpose of this book and following a common use in Latin American Studies, the adjective "American" refers to the Americas as a whole unless otherwise noted. When referring to the United States, the author will use the full name or the acronym US.
23. Their presence in large quantities along the Pacific shores is due to the ocean's geological formation, as the tectonic forces endowed smoldering mountains around the Pacific's edge—the so-called ring of fire— with vast holdings of metals. Dennis O. Flynn, Arturo Giraldez, and James Sobredo, *Studies in Pacific History: Economics, Politics, and Migration* (Burlington, VT: Ashgate, 2002), 1–2.
24. Carrillo, "Los 'chinos' de Nueva España," 18–19; Gordon and Morales, *La plata y el Pacífico*, 80–99.
25. Gordon and Morales, *La plata y el Pacífico*, 86. Mexican silver continued to flow to China well into the twentieth century, long after the galleon had stopped running and the American territories had acquired their independence from Spain. John McMaster, "Aventuras asiáticas del peso mexicano," *Historia Mexicana* 8, no. 3 (January–March 1959): 375. For more on silver, see William Schell Jr., "Silver Symbiosis: Reorienting Mexican

During the first two centuries of colonial administration, Spanish authorities established a rigid mercantilist system that discouraged trade and communication among the colonies in order to protect Spanish merchants. Consequently, a mere handful of ports and maritime routes flourished. The Manila galleons were "part of an empire-wide system of royally sponsored navigation routes or runs (*carreras*), which aimed to regulate trade flows and protect ships from pirates."[26] Two Atlantic fleets circulated every year, leaving together from Seville or Cadiz and separating in Caribbean waters. One fleet continued west towards Veracruz, in New Spain, with stops at a few islands and Yucatecan ports on its way. The other dropped anchor in Venezuelan ports, Cartagena, and finally Portobelo, in the Panamanian isthmus. On the return the fleets met in Havana and sailed back together to Spain.[27] In the Pacific only the Manila galleons were officially sanctioned to circulate once or twice per year.[28] Their cargo was either sold locally at the Acapulco fair, which gathered thousands of people every year, or transferred by land to major cities. The remaining goods were taken to Veracruz in the Atlantic and from there to Spain. At the end of the sixteenth century, the merchants in Callao, Peru, tried to establish a direct run to Manila. After a few successful trips, the crown banned the service to avoid competition for Spanish products. From then on, the only authorized vessels circulating in South America were those carrying silver produced in Peruvian mines. They traveled to Panama, transferring the load by land to Portobelo, on the Caribbean side of the isthmus, and from there to Spain. The Manila galleons mostly carried the silver produced in New Spain's mines across to the Philippines.[29]

While only a few ports were authorized to trade, several others were used to move products around locally, for contraband, or as pirate bases. For example, despite the 1604 prohibition on sailings between the two American viceroyalties and any other contact with Asia apart from the Manila galleons,[30] the so-called naos of Lima continued to circulate between Callao and Puerto Marqués, a few miles north of Acapulco. They also made several stops along the way. Even though there were more active Pacific ports than those authorized by the crown, none except for Manila became a large hub precisely because of their clandestine nature. Even Acapulco, the largest of them all, never materialized into a great city as Europeans preferred to live in cooler places. Only when the galleons arrived did Acapulco come to life. When Humboldt visited the port in the early 1800s, he calculated a

---

Economic History," *Hispanic American Historical Review* 81, no. 1 (2001): 89–133; Gordon and Morales, *La plata y el Pacífico*, 80–99; the various works by Dennis O. Flynn and Arturo Giraldez as well as Mejia, "The Economics of the Manila Galleon," 35–62.

26. Tatiana Seijas, *Asian Slaves in Colonial Mexico: From Chinos to Indians* (New York: Cambridge University Press, 2014), 77.
27. William Lytle Schurz, *El galeón de Manila* (Madrid: Ediciones de Cultura Hispánica, 1992), 335–36.
28. Gordon and Morales, *La plata y el Pacífico*, 49.
29. Schurz, *El galeón de Manila*, 312–13.
30. Gordon and Morales, *La plata y el Pacífico*, 49; Seijas, *Asian Slaves*, 75.

regular population of some 4,000 people, which doubled during the time of the commercial fair.[31]

The Spanish chose Acapulco in the Americas and Manila in Asia as endpoints for the Manila galleon run due to their favorable geographic and geopolitical conditions. Both possessed sheltering bays that protected vessels from external winds and currents and deep waters that facilitated the loading of passengers and merchandise while freeing the Spanish from having to invest in additional infrastructure. From a geopolitical perspective, Acapulco was the closest port to New Spain's capital, Mexico City, and Manila was located at the center of an arch that included Japan, China, India, Malacca, and the Molucca islands, all of which produced highly prized commodities for Spanish markets.[32]

While Spanish merchants monopolized the administration of the Manila galleon, the other central player in the transpacific flux were the Chinese. To begin with, Chinese merchants were the main suppliers of merchandise for the local Philippine colony as well as for the Americas, which they exchanged for the much sought-after silver coins. They arrived in Manila in junks departing primarily from Fujianese ports and carrying an assortment of goods,[33] listed as follows by colonial officer Antonio de Morga in 1609:

> raw silk, . . . numerous velvets, damasks . . . satins . . . and other varied fabrics. They also brought musk, benzoin, ivory; many bed ornaments, curtains, covers and tapestries, . . . tablecloths, cushions and rugs, . . . also some pearls and rubies, sapphires, rock crystals, . . . large quantities of thread of all kinds, sewing needles, . . . even cages with birds that spoke or sung. . . . The Chinese provided an infinite amount of low cost and low value trifles which are much appreciated among the Spanish, as well as many varied pieces of tableware, . . . pepper and other spices, and rarities which, if I had to enumerate each one, would never end even if I could find enough paper for all.[34]

In addition, they participated with local Filipinos in the construction of the large transoceanic galleons built in the Cavite shipyards, just outside Manila.[35] As a result, as many as 30,000 Chinese had settled outside the city walls by the seventeenth century, in a Chinatown known as Parián, in contrast to around 1,000 Spaniards who lived permanently in the city.[36] The impact of Chinese was such

---

31. Schurz, *El galeón de Manila*, 313–17; Gordon and Morales, *La plata y el Pacífico*, 58.
32. Gordon and Morales, *La plata y el Pacífico*, 38–39; Schurz, *El galeón de Manila*, 315.
33. The number of documented Chinese junks visiting Manila increased rapidly, from three in 1572 to around fifty by 1580. Gordon and Morales, *La plata y el Pacífico*, 43.
34. Cited in Gordon and Morales, *La plata y el Pacífico*, 44–45.
35. The Cavite shipyards produced vessels of some 2,000 tons at a time when most big ships were a quarter of this size. With the exception of iron, all the other materials were produced in the Philippines. Gordon and Morales, *La plata y el Pacífico*, 47–48.
36. Gordon and Morales, *La plata y el Pacífico*, 48; Edgar Wickberg, *The Chinese in Philippine Life, 1850–98* (New Haven, CT: Yale University Press, 1965), 6, 21. In his book, Wickberg talks of 20,000 Chinese living in El Parián.

that there were several references to them in daily New Spanish parlance. For example, the Manila galleon was also known as the Chinese nao, the broken pieces of the typical blue and white porcelain coming from China or imitated locally were referred to as "*chinitas*" and served as low-value coins,[37] the road between Acapulco and Mexico City was called the China Road, and Asians who resided in New Spain were all called "Chinos," despite their multiethnic origins.[38]

While the galleon runs following Urdaneta's inaugural sailing were not as disastrous,[39] the voyage continued to be long and extremely challenging. Going westward from Acapulco, the fleet took between forty-five and sixty days to arrive. But eastward from Manila, rough seas and unfavorable winds could slow the return to more than six months, with the added complications of hygienic problems and lack of provisions.[40] As part of his five-year-long trip around the world, Italian traveler Giovanni Francesco Gemelli Careri experienced one of these firsthand as a cabin passenger, writing that it took "204 days and five hours" to complete the journey.[41] His account of the voyage is full of refences to the rough conditions:

> Monday the first of October, . . . the wind came to S.E . . . and at night blew so hard at S. that the pilot was forced to lower his top sails and main yard. A great storm blowing on Tuesday 2d, at S. and the sea beating hard upon us, we were forced to lie by the foresail back'd, and the waves beat so furiously on the rudder, that the whipstaff broke, . . . the ship was toss'd upon vast mountains of water, and then again seem'd to sink to the abyss, the waves breaking over it. No fire could be lighted, and so all ate cold meat, and there was no chocolate to be made . . . and there was no standing or sitting in a place, but we were toss'd from side to side. About midnight I had liked to be knock'd in the head, by two linstocks of the guns falling upon my bed. Wednesday 3d, the same wind continuing, the storm was nothing abated.[42]

He found the trip extremely tedious, "and the more because in a short time all the provisions grew naught, except sweetmeats and chocolate, which are the only comfort of passengers."[43] Wondering why some of the crewmembers and merchants "venture through it four, six, some ten times," he reasoned that it was only due to the profits that some made. "The master, his mate, and boatswain, who may put aboard several bales of goods, may make themselves rich in one voyage."[44] As for himself, after so many hardships, "these nor greater hopes shall not prevail with me

---

37. "Chinitas" means "little Chinas" or "little Chinese." Gordon and Morales, *La plata y el Pacífico*, 57, 62.
38. Seijas, *Asian Slaves*, 1, 9. See Lee, *Making*, 20–29 for an informative summary on Chinos in New Spain.
39. With the exception of fifty-nine galleons that sank, out of the 400 or so trips made in the 250 years that the route lasted. Gordon and Morales, *La plata y el Pacífico*, 51.
40. Gordon and Morales, *La plata y el Pacífico*, 51.
41. Giovanni Francesco Gemelli Careri, *A Voyage to the Philippines* (Manila: Filipiniana Book Guild, 1963), 171.
42. Gemelli, *Voyage*, 149–50.
43. Gemelli, *Voyage*, 156.
44. Gemelli, *Voyage*, 156.

to undertake that voyage again, which is enough to destroy a man, or make him unfit for anything as long as he lives," and he concluded that "the Spaniards, and other geographers, have given this [ocean] the name of the Pacifick sea, as may be seen in the maps; but it does not suit with its tempestuous and dreadful motion, for which it ought rather to be call'd the restless."[45]

If cabin passengers such as Careri found the journey unbearable, those in the lower ranks had an even harsher time. To begin with, the galleons often traveled overloaded to increase profits, so cargo was everywhere, even on deck, leaving little room for common sailors and slaves to move around or find shelter.[46] Mortality was high and corpses were sometimes kept until the ship reached the shore so that they could be buried in sanctified ground, adding to the horrific conditions on board.[47] As of the 1600s, there were around 400 people traveling in a galleon, of whom up to 250 were crew,[48] with only a handful of higher-ranked Spanish sailors and a large majority of lower-ranked Asians. Despite their skill, they could receive half as many rations or less than their Spanish counterparts; they were regularly paid less than promised.[49]

It is calculated that between 40,000 and 60,000, and perhaps up to 100,000, Asians crossed the Pacific in the galleons and settled in New Spain.[50] Almost none were women, especially after the crown ordered officers in Manila to only allow married women on board, following a series of complaints about the few female travelers being treated as slaves and prostitutes.[51] Of the men, at least 8,000 traveled as slaves before this practice was abolished on Pacific runs in 1672.[52] Most of the free and enslaved passengers who settled in New Spain were Chinese, followed by Filipinos and then by a multiplicity of ethnic origins. Many started as sailors who, once ashore, worked in a variety of trades, such as barbers, merchants,

---

45. Gemelli, *Voyage*, 157.
46. Seijas, *Asian Slaves*, 79.
47. Seijas, *Asian Slaves*, 79–80. Seijas describes a galleon in 1629 with a loss of 100 people, but most likely mortality was usually much lower. In his account, Gemelli Careri mentions a handful of sailors who died on his trip, all of whom were thrown overboard. He also described people on board failing ill with scurvy and beriberi. Gemelli, *Voyage*, 171.
48. Seijas, *Asian Slaves*, 78.
49. Lee, *Making*, 21–22.
50. Gordon and Morales, *La plata y el Pacífico*, 50, 73.
51. This crown order was often ignored, however, so that perhaps up to one-quarter of the slaves who crossed the Pacific before slavery was banned were women. Seijas, *Asian Slaves*, 79. Gemelli Careri mentions only three women in his account, of whom only one was aboard his ship and was the captain's wife. He was told of the stories of the other two women by fellow passengers: one drowned and the other was a slave. Gemelli, *Voyage*, 167.
52. Seijas, *Asian Slaves*, 23. After reviewing the official records of the Royal Treasury for Acapulco, Oropeza has calculated an average of thirty-two slaves traveling in each of the 118 galleons that arrived in Acapulco between 1565 and 1673 for a total of 3,776. She also states that it is difficult to know the right number for various reasons, such as corruption and fraud by those involved in the traffic as well as the discrepancy of sources. For instance, in the 1637 records, when a strict review was done, records show the presence of ninety-three slaves in each of the two galleons. If this is a more accurate number, then there could have been up to 10,000 slaves arriving in this period. Oropeza, *Migración asiática*, 148–53.

musicians, copyists, domestic aids, tailors, shoemakers, drivers, and silversmiths.[53] Some formed commercial and personal bonds with locals and a few even became prominent and relatively wealthy members of the towns in which they lived.[54] By the early seventeenth century there were Chinese living in Acapulco and nearby ports, in mining centers close to Tepic and Zacatecas, and in Mexico City.[55]

Despite the frequent contacts and settlements, Spanish colonial records tended to portray Chinese as aliens and often assigned them negative stereotypes. One of the earliest examples is that of the Augustinian bishop of Chiapas, Juan González de Mendoza, who, in a 1585 history treatise, described Chinese as lacking "the clear light of the true Christian religion without which the most subtle and delicate understandings get lost."[56] Similarly, in his *Historia de las guerras civiles de la China y de la conquista de aquel dilatado Imperio por el Tártaro*, Archbishop Juan de Palafox y Mendoza, who briefly served as viceroy in New Spain in 1642, found an explanation for the Tartar conquest of Chinese territories in the latter's "coward," "effeminate," and "servile" personality.[57] Around the same period, in 1635, a group of Spanish barbers from Mexico City complained about their Chinese peers, accusing them of taking clients away from them. They petitioned the authorities to relocate them to the outskirts of the city so that they would not have to face what they considered illegitimate competition from outsiders.[58]

As for Japanese, records show that their contacts with New Spain were scarcer. The route that the galleons followed circulated near the coasts of Cipango—the name given by Spaniards to present-day Japan—where vessels stopped when in danger. The most famous of these was *San Francisco*, which shipwrecked off the coast of Kazusa en route to Acapulco in September 1609. Among its 300 passengers was former governor of Manila Rodrigo de Vivero, who had had epistolary contact with the shogun during his brief interim government in 1608 and favored exchanges between territories. The local authorities warmly welcomed him to Kazusa and offered him a new ship, renamed *San Buenaventura*, and a loan to cover unforeseen expenses. After a few months Vivero departed for Acapulco with twenty-three Japanese merchants aboard *San Buenaventura*.[59] This episode has long

---

53. Gordon and Morales, *La plata y el Pacífico*, 74; Schurz, *El galeón de Manila*, 98; Anthony Reid, ed., *The Chinese Diaspora in the Pacific* (Burlington, VT: Ashgate Variorum, 2008), xxi–xxiii. For the role played by Chinese in the Philippines during the era of the galleons, see Wickberg, *Chinese in Philippine*, 3–43.
54. See Oropeza, *Migración asiática*, 165–292.
55. Xu Shicheng, "Los chinos a lo largo de la historia de México," in *China y México: Implicaciones de una nueva relación*, ed. Enrique Dussel Peters and Yolanda Trápaga Delfín (Mexico City: La Jornada Ediciones, 2007), 53–54; Evelyn Hu-DeHart, "Mexico," in *The Encyclopedia of the Chinese Overseas*, ed. Lynn Pann (Cambridge, MA: Harvard University Press, 1998), 256; Leo M. Dambourges Jacques, "The Chinese Massacre in Torreon (Coahuila) in 1911," *Arizona and the West* 16 (Autumn 1974): 234; Schurz, *El galeón de Manila*, 98; Reid, *Chinese Diaspora*, xxi–xxiii.
56. Cited in José Jorge Gómez Izquierdo, *El movimiento antichino en México (1871–1934): problemas del racismo y del nacionalismo durante la Revolución Mexicana* (Mexico City: INAH, 1991), 18.
57. Gómez, *Movimiento antichino*, 19–20.
58. Hu-DeHart, "Mexico," 256.
59. Enrique Cortés, *Relaciones entre México y Japón durante el Porfiriato* (Mexico City: SRE, 1980), 23–24.

been regarded by both the Mexican and Japanese governments as the symbolic beginning of bilateral relations.[60] While the first official exchanges between Spanish and Japanese had been mostly courteous, by the 1630s they were eroded through increasing mutual mistrust. On the one hand, Spain's Manila-based merchants did not want to face possible competition from Japanese traders, so they urged that relations be confined to a minimum. On the other hand, Japanese authorities had become suspicious of the European influence within their territories, particularly that of Spanish missionaries, and began restricting exchanges.[61] Consequently, contact between Japan and the Spanish American territories in the following two and a half centuries remained limited and fortuitous, even accidental, comprised mostly of shipwrecks on both coasts.[62]

The Spanish crown abolished many of its trade restrictions in the mid-eighteenth century as part of a series of measures commonly known as the Bourbon reforms, in reference to the royal dynasty governing Spain since 1713. In an effort to regulate and tax the empire's excessive trade in contraband, Charles III signed the October 1778 Reglamento de Comercio Libre, permitting inter-provincial exchanges and opening twenty-two American ports to trade.[63] The Pacific ports benefiting from the measure were Valparaíso and Concepción in Chile, and Arica, Callao, and Guayaquil in the viceroyalty of Peru.[64] San Blas, north of Acapulco, also grew as it became a stopover or terminus for the galleons,[65] as well as the base of support for Spanish coastal explorations and for the missionaries sent to colonize and catechize the northern provinces of Sonora and the Californias.[66]

---

60. In the fall of 2009, the governments of Mexico and Japan celebrated what they labeled as the 400th anniversary of relations between the two nations with numerous festivities in both countries.
61. Archivo Histórico Genaro Estrada-Secretaría de Relaciones Exteriores (AHGE-SRE), L-E-2259, 1–7; Cortés, *Relaciones*, 11–31.
62. One of the best documented cases is that of *Eijyu-maru*, a Japanese vessel that shipwrecked in 1841. Seven of its crew members ended up landing on the southwestern coast of the Baja California peninsula. See Kiyoshi Irie, "Nuevos relatos sobre México," *Estudios de Asia y África* 92, no. 3 (September–December 1993): 421–48. Irie's original work in Japanese is entitled *Meshiko Shinwa* and was published in 1964.
63. Schurz, *El galeón de Manila*, 333.
64. *Reglamento y Aranceles Reales para el Comercio Libre de España a Indias de 12 de Octubre de 1778* (Madrid: 1778), 9. On Bourbon economic reforms, see Kenneth J. Andrien and Lyman L. Johnson, *The Political Economy of Spanish America in the Age of Revolution, 1750–1850* (Albuquerque: University of New Mexico Press, 1994); Richard L. Garner, *Economic Growth and Change in Bourbon Mexico* (Gainesville: University Press of Florida, 1993); Jordana Dym and Christophe Belaubre, eds., *Politics, Economy, and Society in Bourbon Central America, 1759–1821* (Boulder: University Press of Colorado, 2007). In the second part of his book, Bonialian discusses the repercussions of the Bourbon reforms in the Pacific. Bonialian, *América española*, 201–366.
65. AGI, Fondo Ultramar, Sección Quinta Audiencia de Filipinas, 688. The documents dated between 1781 and 1783 make at least five references to vessels leaving San Blas for the Philippines.
66. Walton Bean and James J. Rawls, *California: An Interpretive History* (New York: McGraw Hill, 1988), 20–21. See also Guadalupe Pinzón Ríos, "Acciones y reacciones en los puertos del Mar del Sur. Desarrollo portuario del Pacífico novohispano a partir de sus políticas defensivas, 1713–1789" (PhD diss., Facultad de Filosofía y Letras, UNAM, 2008), 197–202.

The Spanish living in the Philippines also diversified their maritime relations during the period of the Bourbon reforms. This process began in 1785, after the Spanish king decreed the foundation of the Royal Philippine Company (RPC) to establish direct trade with Asia, Spain, and South America. The company's first vessel sailed in October from Cadiz to Manila by way of Cape Horn, with a stop in Callao (Lima's port). Soon, two other ships followed by way of the Cape of Good Hope. Between its foundation and 1813, RPC sent over sixty vessels to different ports in Asia and Spanish America.[67] In addition, the decree in 1789 authorizing all European vessels carrying goods to and from Asian ports to enter Manila also contributed to the archipelago's commercial opening.[68] This liberalization of commerce as well as the diversification of Fujianese interests to varied parts of Southeast Asia caused the junk trade to decrease substantially during the eighteenth century. In effect, between 1797 and 1812 there was an average of only eight junks arriving annually at Manila. Just as the Philippines' Fujianese links began to fade, ties with Canton increased as RPC vessels were restricted from trading anywhere other than there. The maritime connections between Philippines and Cantonese ports, notably Canton and Hong Kong after the British colonial government started in 1842, would only grow in the following decades.[69]

The Bourbon reforms did little to stop the decay of the Spanish crown's influence over the Americas. At the turn of the century, and particularly after Napoleon's invasion of Spain in 1808, a series of revolts spread throughout the kingdom. In 1811, the galleon *Magallanes* arrived in Acapulco only to have its cargo seized by local rebels in the context of the revolution of independence against the Spanish authorities. Unable to guarantee the safety of the galleon and with little control over Acapulco and New Spain, the crown decreed the suppression of the line in October 1813. Two years later *Magallanes* sailed for Manila for the last time. The voyage officially ended the 250-year-long route and simultaneously symbolized the decline of Spanish supremacy over transpacific exchanges.[70]

---

67. Nicholas P. Cushner, *Spain in the Philippines: From Conquest to Revolution* (Manila: Ateneo de Manila University, 1971), 190–91.
68. Prior to this only Spanish galleons and Chinese junks were legally allowed to dock and trade in the Philippines.
69. Wickberg, *Chinese in Philippine*, 21–22. See his chapter on foreign trade for the increasing connections with Canton.
70. Schurz, *El galeón de Manila*, 60. There were a few more galleons that circulated up to 1821, most of them landing in the port of San Blas. Yet the last one to do so officially was *Magallanes*. Vera Valdés Lakowsky, *Vinculaciones sino-mexicanas: albores y testimonios, 1874–1899* (Mexico City: UNAM, 1981), 62. Bonialian offers what he calls a "structural explanation" for the decay of the Manila galleons since the eighteenth century, linked to the Bourbon reforms, which includes aspects such as the opening and flourishing of new trading routes around Cape Horn; the presence of Spanish merchants who competed with the galleons using their own personal ships; and the "atlantization" of Asian trade, that is, the introduction of Asian goods to New Spain via Atlantic ports, among others. Bonilian, *América española*, 206–22.

The Steam Revolution and the Pacific Imperial Projects of Great Britain and the United States

Between 1822 and 1823, a frustrated bureaucrat from Manila's post office complained repeatedly to his superiors in Spain of delivery service disruptions. In his first letter, dated March 22, 1822, he lamented the "complete disorder in which official and personal mail arrived via the Spanish brigantine merchant ship *Ruperto* on its trip back from the Port of San Blas." In February 1823 he reported that two boxes of mail sent to Spain close to two years earlier via San Blas, Mexico, had just been returned to Manila. He announced his decision to resend them in a British frigate, *Mermaid*, that was not stopping in the Americas. He also requested permission to communicate "directly with the General Management [in Spain] without having to report to the Mexican Administration on which we used to depend in the past." Ten months later he accused the Mexican government of holding on to most of the mail due for Manila and asking for exorbitant fees to release it. He urged authorities "to change the way to send [mail] from what has been done previously" so that New Spanish ports, no longer under the crown's administration, stopped being the intermediary.[71]

When the American colonies of Spain gained independence in the 1820s,[72] communications in the Pacific, as demonstrated by the post officer's complaints, were altered for good. The system of runs created by the Spanish Empire that had begun in the late sixteenth century, centered on the Spanish galleons going from Manila to Acapulco and later, in the eighteenth century, also to Callao, Valparaiso, and San Blas, were quickly terminated. This prompted a diversification of routes and a proliferation of vessels from various origins circulating in the Pacific. For instance, according to Manila's Commercial Registry of 1829, of the 110 vessels that docked from abroad throughout the year, only thirty-two were of Spanish origin. The rest belonged to nine other nationalities, with the largest number—twenty-nine—being from the United States, followed by fifteen from England. As for the 113 vessels that left Manila for international ports, the pattern was similar: thirty-six ships were from Spain, twenty-nine from the United States, seventeen from England, and the remaining thirty-one originated from nine other nationalities. Of all the inbound runs, only six came directly from the Americas: five from Callao and one from Valparaiso; outbound, there was a sole ship crossing to Valparaiso. None came from or left for Acapulco or any other ports in the former New Spain, with the most frequent destinations being Macao, Lingding Island, and Singapore. While Manila, still a Spanish port up to 1898, continued to be frequently connected to Chinese ports as in previous centuries, the trips were no longer made primarily by Spanish

---

71. Letters from March 22, 1822, February 13, 1823, and December 12, 1823, found in AGI, Fondo Ultramar, Legajo 644, Sección 10a, Ministerio de Ultramar.
72. With the exception of Cuba and Puerto Rico, all the continental Spanish American colonies obtained their independence from Spain in the 1820s.

and Chinese boats but rather by those from the United States and England.[73] In the case of independent Mexico, the port of Mazatlan replaced Acapulco as the endpoint of the now intermittent direct traffic going to Asia, particularly to Hong Kong after 1842—the year the British took over its administration—and from there to the Chinese ports in Amoy, Foochow, Ningpo, and Shanghai, which had recently been opened to foreign trade. Other, much longer routes from Mexico to Asia departed either from Mazatlan, San Blas, or Guaymas in the Pacific or from Veracruz on the opposite coast to London and then Asia via the British sailing and later steamer fleet navigating the Atlantic and Indian oceans. They mostly carried silver coins produced in Mexican mines to China,[74] albeit in smaller quantities compared with colonial times, in exchange for British imported cotton and other goods.[75]

Just as the history of transpacific sailing vessels remains linked to the role played by Spanish imperialism, the development of steam navigation in the Pacific came hand in hand with the pre-eminence of British imperial interests in the nineteenth century. Situated at the vanguard of the Industrial Revolution, the British were the first to conceive, finance, and patent many of the inventions related to steam-powered navigation.[76] In this context, the British introduced the first ocean

---

73. All the stats were determined by the author based on the Manila registries found from January to November 1829 (there were no stats listed for December) in AGI, Fondo Ultramar, Legajo 664, Registro Mercantil de Manila de enero hasta noviembre de 1829. The nine other nationalities were Chinese, Danish, Dutch, French, Hamburg, Portuguese, Prussian, Russian, and Sandwich.
74. See letter from January 30, 1834, where the local post officer informs Spain that there are several types of silver coins circulating, some from the "dissident American countries," some still with the old Spanish seal. AGI, Fondo Ultramar, Legajo 644, Sección 10a, Ministerio de Ultramar.
75. Bonialian, *América española*, 321–30; Schell, "Silver Symbiosis," 108–17. On page 110 Schell also states that prior to 1850 Mexico imported around 100,000 pounds of Chinese raw silk every year. See the entire article to understand the cycles of silver exchanges. As we will see later in this chapter, in the 1850s, the most popular route between Mexican and Asian ports, notably Hong Kong, was via San Francisco. Elizabeth Sinn, *Pacific Crossing: California Gold, Chinese Migration, and the Making of Hong Kong* (Hong Kong: Hong Kong University, 2013), 176; Busto, *Pacífico mexicano*, 190–92.
76. To begin with, the steam engine, developed by the Scottish engineer James Watt between 1769 and 1782, became the basis for steamers. At the time, the Forth and Clyde Canal, which crosses Scotland, became the trial zone for a series of steamboats invented at the turn of the nineteenth century, culminating in the building of the *Charlotte Dundas*, the first practical steamboat, developed by William Symington, as well as Henry Bell's *Comet*, the first satisfactory paddle steamer offering local regular transport as of 1812. The shipyards at London, Liverpool, and Glasgow were the most prolific at fabricating paddle wheelers and screw steamers with wooden hulls, and later, in the 1860s, they were the first to change over to iron vessels, better suited to endure the strain caused by the pitching and consequent racing of the engines of a screw steamship. A screw steamship—as opposed to a paddle wheeler—transported itself by means of a propeller. Will Lawson, *Pacific Steamers* (Glasgow: Brown, Son & Ferguson, Ltd., 1927), xiii–xv, 20–21; Roland E. Duncan, "William Wheelright and Early Steam Navigation in the Pacific 1820–1840," *The Americas* 32, no. 2 (October 1975): 257. On the relationship between British imperialism and the development of the steamship industry, see David Kilingray et al., *Maritime Empires: British Imperial Maritime Trade in the Nineteenth Century* (Woodbridge: Boydell Press/National Maritime Museum, 2004); Ben Marsden and Crosbie Smith, *Engineering Empires: A Cultural History of Technology in Nineteenth-Century Britain* (New York: Palgrave Macmillan, 2005); chap. 6 of Ian Friel, *Maritime History of Britain and Ireland c. 400–2001* (London: British Museum Press, 2003); *Shipping, Technology, and Imperialism: Papers Presented to the Third British–Dutch Maritime History Conference* (Brookfield, VT: Ashgate, 1996). For a general overview on the development of the steamship industry, see Thomas Crump, *The Age of Steam: The Power That Drove the Industrial Revolution* (London: Constable & Robinson, 2007): 284–320.

steamers to the Pacific as early as the 1820s. All were short-lived ventures but they set the precedent for the first liners that would emerge decades later. *Rising Star*, the first of them all, arrived in Valparaiso, Chile, in May 1822. Even though it mostly used wind rather than steam to navigate, it became a symbol of power and prestige when compared with its contemporary sailing vessels, as is evident in the remarks of Lady Calcott, one of its British passengers: "it was no small delight that I set my foot on the deck of the first steam-vessel that ever navigated the Pacific, . . . our stately vessel glided smoothly and swiftly through them [sailing vessels] without a sail, against wind and waves, carrying on its deck a stronger artillery than Almagro ever commanded."[77] *Rising Star* soon returned to England as the Chilean government was unable to pay for it. *Telica* arrived next from Liverpool to Guayaquil, Gran Colombia, in August 1825, with the intention of traveling regularly to and from Callao, but an on-board explosion destroyed it. The third steamer and first in North America, *Beaver*, operated around Victoria, British Columbia, in the late 1830s.[78]

Crossing the Pacific propelled by steam was no simple task, so navigation remained littoral until mid-century. In contrast, steamships traversed the Atlantic starting in the 1820s and regular liners plied the route from the late 1830s.[79] The reasons for the delay were multiple. The first was a question of size: the Pacific is twice as wide as the Atlantic. The task, therefore, required double the amount of coal. With no supplies available along the way, the odds for success were substantially lowered. Additionally, the shortest route crossed through the heart of one of the ocean's stormiest sections, particularly during the winter season. Inclement weather could delay a trip for days, exhausting crucial fuel supplies, and there were no safe harbors along the way for repairs. Furthermore, establishing a dependable steamship business required substantial capital investment. Prior to the late 1860s, neither government nor private enterprise had amassed sufficient economic, political, and technological clout to establish a regular transpacific steamer run.[80]

The first steamship company offering regular cargo and passenger runs in the Pacific was British. Formed in Liverpool, with a starting capital of £250,000 composed partially of governmental subsidies, Pacific Steam Navigation Company (PSN) was granted a royal charter by Queen Victoria in February 1840. London's Curling, Young & Co. was assigned to build two brand-new paddle steamers for the run, *Chile* and *Peru*, which had cabin accommodations for 150 passengers and a freight capacity of about 300 tons.[81] In October the South American Valparaiso-to-Callao line was inaugurated, "with a worthy reception, composed of the military bands of Valparaiso. . . . Both steamers criss-crossed the bay in different directions,

---

77. Diego de Almagro was a Spanish conquistador who participated in the military conquest of Peru in the early sixteenth century and led the first European expedition through the territories that we know today as Chile. Cited in Duncan, "William Wheelright," 259.
78. Duncan, "William Wheelright," 261, 270.
79. Crump, *Age of Steam*, 286–93.
80. Tate, *Transpacific Steam*, 21–24; Lawson, *Pacific Steamers*, xiii, xix.
81. Duncan, "William Wheelright," 276–81; Lawson, *Pacific Steamers*, 2–4.

receiving the salutations of the multitude attracted by the novel spectacle, and anchored near the shore."[82] Around the same time, other British companies also started local transports on the Australian and Chinese coasts. By the early 1850s, the British had established a regular travel schedule to and from England, Australia, and China by way of the Indian Ocean.[83]

Two other factors that contributed to the acceleration of steam navigation by mid-century were the discovery of large deposits of gold in different coastal areas and the expansion of the United States into the Pacific.[84] In the case of the former, starting in 1848, a series of so-called gold rushes in California,[85] Australia, northwestern Canada, New Zealand, and Alaska sparked the movement of large numbers of people. The fastest way to move them was by sea. With such a large demand for maritime transportation, regular passenger service began via steamer and the ports nearest to the gold rushes became increasingly urbanized. In the case of the latter factor, in the 1840s, the US government secured a large Pacific coastline through war and diplomacy, including by settling a border dispute with the British over the so-called Oregon territory. In June 1846, after close to three decades of joint administration, both governments agreed to respect the forty-ninth parallel as the legitimate border between the two countries. This granted the United States the exclusive administration of the coastline north of Mexico, against which the US Congress had declared war barely a month earlier. After two years of military confrontation, the Mexican authorities capitulated. In February 1848, representatives of both countries signed the Guadalupe–Hidalgo treaties, which stipulated that the provinces of Alta California and Santa Fe de Nuevo México would form part of the United States. In only two years, the United States gained control of some 2,500 kilometers of Pacific coastline.[86]

In order to establish contact with the newly acquired territories, on March 1847, Congress requested the Secretary of the Navy to call for bids for a mail service between New York and Chagres, Panama, and another between Panama City and Astoria, Oregon.[87] William H. Aspinwall, a New York businessman, won the bid and the subsidy for the Pacific route, creating PMSS in April 1848. It became the first US company offering regular service on the Pacific coast. *Oregon*, *California*, and *Panama*, the three new 1,000-ton side-wheelers with accommodations for seventy-five passengers including cabin and steerage, built in New England's shipyards, were ready to sail by the fall of 1848. As per usual at the time, they also possessed

---

82. Cited in Duncan, "William Wheelright," 281 from the *Mercurio*, a Chilean newspaper. The celebration was memorable enough to be re-enacted on the centennial in 1940.
83. Lawson, *Pacific Steamers*, 5, 13.
84. Lawson, *Pacific Steamers*, 18.
85. Busto, *Pacífico mexicano*, 23.
86. Donald D. Johnson, *The United States in the Pacific: Private Interests and Public Policies, 1784–1799* (Westport, CT: Praeger, 1995), 78.
87. John H. Kemble, "Pacific Mail Service between Panama and San Francisco, 1849–1851," *Pacific Historical Review* 2, no. 4 (December 1933): 406.

sails. *California* left New York for Panama City via Cape Horn on October 6, 1848.[88] While the contract specified that mail cargo would be a PMSS priority in exchange for governmental subsidies, it soon became evident that the transportation of gold-seekers was good business. When *California* reached Panama on January 17, 1849, over 1,500 people wanted to board. But *California* was already full, having already picked up a group of Peruvians at the port of Callao. The chaos took days to settle. In the end, company officials gave preference to those who had purchased tickets directly from the New York branch. The Peruvians ended up in improvised standee berths and a total of 365 passengers—in a vessel that had accommodations for only seventy-five—left for California. It took the vessel twenty-eight days to arrive at its destination. Within the next two decades, steamers would shave travel time down to thirteen days and San Francisco would become the route's terminal. This chaotic voyage inaugurated the monthly and later semi-monthly Panama to Oregon run.[89]

The first transpacific steamers appeared by mid-century, once again motivated by a gold frenzy. *Monumental City* and *New Orleans*, the first two transpacific steamships, for example, left San Francisco for Sydney, both gold rush destinations, in February and March 1853 respectively. They took nine weeks to make the crossing and brought hundreds of gold-seekers to Australia's coast. The third steamship to cross the Pacific came from the opposite direction. The suitably named *Golden Age* originally left New York for London in September 1853. It then traveled to Melbourne via the Indian Ocean, where it did a few local runs. Finally, on May 11, 1854, it took passengers, cargo, and mail to Panama, being the first steamer to cross the Pacific on a proposed regular basis. In the end, a dependable steamship link between Sydney and Panama did not materialize until 1863, when the Panama, New Zealand & Australian Royal Mail Company was formed, subsidized by both the New Zealand and British governments.[90] Other transpacific steamers traveling between San Francisco and Hong Kong in the period 1862 to 1864 were Prussia's *Scotland*, Britain's *Robert Low*, and *Oriflamme* from the United States.[91]

The United States' acquisition of the California and Oregon territories implied not only the consolidation of a coast-to-coast national project, but also the beginning of the nation's imperial strategy in the Pacific.[92] This provided an incentive for transpacific naval expeditions, which soon incorporated the use of steamers. For instance, the 1853 and 1854 fleets led by Commodore Matthew Perry to Japan to obtain concessions for the United States included at least three steamers—*Mississippi*, *Susquehanna*, and *Powhatan*—of around 2,000 tons.[93] They served

88. Kemble, "Pacific Mail," 406–7.
89. Kemble, "Pacific Mail," 412, 417.
90. Lawson, *Pacific Steamers*, 7–9.
91. Tate, *Transpacific Steam*, 23.
92. Arthur Power Dudden, *The American Pacific: From the Old China Trade to the Present* (Oxford: Oxford University Press, 1992), xix.
93. Arthur Walworth, *Black Ships Off Japan: The Story of Commodore Perry's Expedition* (New York: Alfred A. Knopf, 1946), 239.

not only as rapid transportation but also as a display of power and technological advancement. In effect, while steamships were known to a few Japanese sailors and fishermen, no one had yet seen a steamer navigate the Uraga and Edo bays, so close to the empire's capital. Many locals were surprised by the smoke and the noises coming out of Perry's ships as well as by their capacity to run fast against the wind. Additionally, steamers played a central role in the demands made by the US government to Japan, as shown in the letter sent by President Millard Fillmore via Commodore Perry in March 1853: "we understand there is a great abundance of coal and provisions in the Empire of Japan. Our steamships, in crossing the great ocean, burn a great deal of coal, and it is not convenient to bring it all the way from America. We wish that our steamships and other vessels should be allowed to stop in Japan and supply themselves with coal, provisions, and water."[94] The shogun acquiesced in March 1854, permitting the United States' vessels to land, trade, and replenish coal and other supplies in Japanese ports.[95] Subsequently, *Susquehanna* and *Mississippi* returned to Honolulu and then continued around South America to become the first US steam warships to circumnavigate the globe, an impressive display of the nation's maritime power.[96]

The British and French navies also used steamers in imperialist ventures in Asia. For instance, between 1854 and 1855 both countries sent two joint expeditionary forces to Siberia in the context of the Crimean War. Each squadron had a steam-assisted vessel, both of British manufacture: *Virago* in 1854 and *Brisk* a year later. Their transpacific journey started in the port of Callao, Peru. A few decades earlier, after the first Opium War of 1839–1842, the British navy had introduced several steamers to the coasts of China, but they had not traversed the Pacific and rather arrived by way of the Indian Ocean.[97]

By the early 1860s, steam had become a technology regularly used in Pacific navigation, although always in conjunction with sails. While several steamers, both war and merchant ships, had crossed the ocean, there was only one regular transpacific service running, that of the Panama, New Zealand & Australian Royal Mail Company, between Panama and Sydney, initiated in 1863 with British subsidies. British investors also controlled local and regional steamship routes on both sides of the Pacific, particularly on the coasts of Australia, China, and South America.

---

94. "Letter of Millard Fillmore, President of the United States of America, to his Imperial Majesty the Emperor of Japan," in Walworth, *Black Ships*, 250.
95. See Julia H. Macleod et al., "Three Letters Relating to the Perry Expedition to Japan," *The Huntington Library Quarterly* 6, no. 2 (February 1943): 229; Louise P. Kellogg, "The United States and Japan," *The Wisconsin Magazine of History* 4, no. 3 (March 1921): 347–49; Arthur J. Marder, "From Jimmu Tennō to Perry: Sea Power in Early Japanese History," *The American Historical Review* 51, no. 1 (October 1945): 32; George Feifer, *Breaking Open Japan: Commodore Perry, Lord Abe, and American Imperialism in 1853* (New York: Smithsonian/HarperCollins, 2006): 4–6. For a detailed account on Perry's expeditions to Japan, see Walworth, *Black Ships*; Feifer, *Breaking Open Japan*. More on this topic in Chapter 4.
96. Tate, *Transpacific Steam*, 21. According to Tate, the other warships from the United States that crossed the ocean westward before 1865 were *Saginaw* in 1860 and *Wyoming* two years later.
97. Tate, *Transpacific Steam*, 21.

In North America, PMSS's Panama to San Francisco route eclipsed all others in importance. Mexico's location in-between these ports influenced the way its coastline became incorporated into the dynamics of the Pacific steamship services.

The Mexican Ports and the San Francisco–Panama Axis[98]

While San Francisco and Panama had storied maritime traditions, it was not until the mid-nineteenth century that they became the main international Pacific ports of the Americas. San Francisco had been used by the Spaniards since the late eighteenth century but remained a relatively small port for local transportation and for the replenishment of sailing vessels participating in the traffic of animal skins, particularly of sea otters, seals, and beavers, that were sold in China in exchange for porcelain, silk, tea, and other commodities. The China trade was an important incentive for the United States to seek control of California during the mid-century war with Mexico described earlier.[99] Yet it was the 1848 gold rush that made San Francisco boom. Its large, protected bay, capable of sheltering hundreds of ships at a time, soon became the preferred docking option. The rapid construction of wharfs as well as the arrival of increasing numbers of ships exemplify the fast pace at which the port grew during those years. For instance, the first private stone pier was built in 1847, while the first public one was constructed in September 1848; within a few years, every single street ended in a pier. The gold rush reshaped the city—Yerba Buena Cove was landfilled to build more wharves and to accommodate the expanding city center. In terms of the quantity of ships docking there, in July 1849 some 500 vessels anchored in the bay. Two years later, the number had increased to 800.[100] Panama, meanwhile, had a much longer international maritime tradition dating at least to the sixteenth century. Yet the port's traffic had diminished considerably since the eighteenth-century Bourbon reforms; RPC inaugurated routes along Cape Horn, around South America, without touching Panama.[101] However, it was in the gold-rush context that Panama consolidated as the most transited port of the area as

---

98. While several authors had talked about the San Francisco–Panama route since the 1930s, Karina Busto Ibarra proposed the notion of the San Francisco–Panama geohistorical axis. "Axis" refers to "the idea that spaces articulate through interconnected centers or localities, whose transformations have global repercussions." Busto, *Pacífico mexicano*, 24. She explains the axis throughout her book, but especially in chap. 3: 131–58. Bonialian defines geohistorical axes as "supra-regional structures with little change . . . that combine maritime with land routes where people, commodities and cultural elements circulate." Bonialian, *América Española*, 20.
99. See Jim Hardee, "Soft Gold: Animal Skins and the Early Economy of California," in *Studies in Pacific History: Economics, Politics, and Migration*, ed. Dennis O. Flynn, Arturo Giraldez, and James Sobredo (Burlington, VT: Ashgate, 2002), 23–39; Sinn, *Pacific Crossing*, 37.
100. Tate, *Transpacific Steam*, 17–19. For a study of San Francisco and its exchanges with Mexican ports, see Busto, *Pacífico mexicano*, 161–98.
101. Dení Trejo Barajas, "Del Mar Caribe al Mar del Sur. Comercio marítimo por el Pacífico mexicano durante las guerras de independencia," in *Entre la tradición y la modernidad. Estudios sobre la independencia*, ed. Moisés Guzmán (Morelia, MI: UMSNH, 2006), 353. In the article the author suggests that during the War of Independence, Panama's trade recovered for a short period.

it offered the shortest journey across the American continent as well as an adequate land and maritime infrastructure, sponsored by British and US investors. In 1850, the Panamanian route surpassed that of Cape Horn as the preferred way to travel from the Atlantic coast of the United States to California. During the period 1850 to 1869—the date of the completion of the transcontinental railroad in the United States—between 15,000 to 20,000 people used the Panamanian isthmus to cross from the Atlantic to San Francisco every year. The completion of the Panamanian transcontinental railroad, the first in the Americas, in 1855 helped position Panama as the favored route for gold-seekers and travelers coming from the east coast of the United States and from the Atlantic in general.[102]

Given the technology of the 1850s, steamers could not make the three- to four-week trip between Panama and San Francisco without stops as there was not enough space on board to carry all the necessary supplies, particularly coal. The Mexican coastline thus offered a midway point where PMSS vessels and other freighters could reload food, water, and fuel. The nation's coast also produced enough provisions not only for the travelers but also for the booming population of San Francisco. Indeed, during the first years of the gold rush, San Francisco imported all sorts of foodstuffs, clothes, cigars, and alcohol from Mexico.[103] In this context, the regular PMSS maritime flow allowed several Mexican ports to prosper, particularly Acapulco and Mazatlan, incorporating them into the international steamship circuit of Pacific exchanges as stopovers on the San Francisco–Panama run.[104]

In the case of Acapulco, as of the 1820s, it was a small port used mainly for local voyages and irregular longer trips; once the gold rush started, an increasing number of vessels stopped there to amass supplies of food and drink. Some passengers even overnighted in the various private homes and hotels. Additionally, the mail and travelers coming by land from different parts of the country, notably Mexico City and Veracruz, used Acapulco as a stopover where they could easily find transportation to other Pacific ports. Furthermore, PMSS established a coal station and permanent offices there as early as 1849. Until 1872, when the Mexican government granted a subsidy to PMSS so that its steamers would stop at other Mexican ports, Acapulco remained Mexico's main harbor and every single one of its vessels visited it.[105]

In contrast to Acapulco, Mazatlan boomed after independence. In an effort to regulate illegal maritime traffic, in 1822 it became one of the ports that the Mexican

---

102. John Haskell Kemble, "The Gold Rush by Panama, 1848-1851," *The Pacific Historical Review* 18, no. 1 (February 1949): 45–46. See also Aims McGuinness, *Path of Empire: Latin American Transformations and the California Gold Rush, 1848-1856* (Ithaca, NY: Cornell University Press, 2008) and Busto, *Pacífico mexicano*, 110–22.
103. See Inés Herrera, "Comercio y comerciantes de la costa del Pacífico mexicano a mediados del siglo XIX," *Historias* 20 (April–September 1988): 129–35.
104. Busto, *Pacífico mexicano*, 141.
105. Busto, *Pacífico mexicano*, 249–64.

government opened to international trade.¹⁰⁶ During colonial times Mazatlan had remained a small village whose sheltering bay was used for local exchanges and by pirates and merchants involved in pearl and precious metal contraband. But after 1822, French, British, and especially Germans began to settle and opened businesses to profit from the port's proximity to several gold and silver mining centers.¹⁰⁷ These businessmen were part of a diaspora of Europeans looking for places to invest the capital surpluses generated in the context of their hometowns' booming industrialization. By 1845 these foreigners owned thirteen major commercial houses dedicated to the import of textiles, wines and liquors, furniture, industrial machinery, and hardware.¹⁰⁸ They also exported gold and silver, followed by brazilwood, animal skins, sugar, and pearls. With this flurry of economic activity, Mazatlan briefly emerged as North America's most prosperous port, to be overtaken by San Francisco at the time of the California gold rush.¹⁰⁹ When PMSS started its regular schedule between Panama and Oregon in 1848, Mazatlan became an obvious stopover as it already possessed the necessary shipping infrastructure.¹¹⁰

With time, not only Acapulco and Mazatlan, but also San Benito, Salina Cruz, Manzanillo, and San Blas became stopovers for PMSS vessels. Of these, Manzanillo and Salina Cruz,¹¹¹ as we will see in Chapter 5, would turn into international maritime hubs after they became railway terminals in 1889 and 1907 respectively. The

---

106. In 1820, still during the War of Independence, the Spanish government decreed the opening of Mazatlan to international trade, yet, in reality, it was not until the war ended and the new Mexican government took over that the authorities finally opened a customs office. Rigoberto Arturo Román Alarcón, *El comercio en Sinaloa, siglo XIX* (Culiacán: DIFOCUR/FOECA/CONACULTA, 1998), 16–17.
107. It was closer for Europeans to settle in the Gulf of Mexico and the Caribbean coasts, yet the fact that the customs duties were lower on the Pacific coast offered an additional incentive for foreigners to settle and invest their money on the Pacific coast. Román, *Comercio en Sinaloa*, 19.
108. Six Germans, three French, one Swiss-Spanish, one French-Spanish, one Anglo-Philippine, and one from the United States. Román, *Comercio en Sinaloa*, 19.
109. Román, *Comercio en Sinaloa*, 18, 44–46.
110. For an analysis of the evolution of Mazatlan and its role in the San Francisco–Panama axis, see Busto, *Pacífico mexicano*, 199–248.
111. Since the early sixteenth century, the Spanish conquistadores, travelers, and pirates had regularly used Manzanillo's twin bays—then baptized as Salagua and Santiago de Buena Esperanza. Indigenous locals had been using them for decades. In the following two centuries, the galleons visited them regularly to obtain food supplies from the nearby haciendas specialized in the production of citrus fruits, salt, coconut trees, and cattle. After Independence, the Mexican government opened the port to foreign trade in 1825. But the Mazatlan merchants, seeing this as a threat to their power, lobbied the federal government to revoke the order. As a consequence, in 1837, Manzanillo was only allowed to deal with internal maritime traffic and five years later its port was closed down. When US troops invaded Mazatlan in 1847, maritime traffic was diverted in part to Manzanillo and it slowly transformed into a larger seaport, particularly after the construction of the railroad linking it to Colima, the state capital, in 1889. See Héctor Porfirio Ochoa Rodríguez, "Manzanillo: el intrincado despertar de un puerto," in *Los puertos noroccidentales de México*, ed. Jaime Olveda and Juan Carlos Reyes (Zapopan: El Colegio de Jalisco, 1994): 113–19; Pablo Serrano Álvarez, "Comentario," in *Los puertos noroccidentales*, 66–70. See Chapter 5 for the history of the port of Salina Cruz. The flourishing of both San Benito and San Blas was linked to the export crops produced in nearby haciendas. In the case of the latter, since the late eighteenth century it had become an important port of trade with California in the context of the Bourbon reforms, as mentioned earlier in this chapter. For more information on these ports and the roles they played in the San Francisco–Panama axis, see Busto, *Pacífico mexicano*, 310–13, 336–41, 364–68.

experience, capital, and prestige gained by the San Francisco–Panama steamers stopping in all these Mexican ports allowed PMSS to open a second route, from San Francisco to Asia.

Colorado and Transpacific Relations between Mexico and Asia in the 1860s

On the first day of 1867, San Franciscans woke up to a street festival. Enjoying a respite from one of the rainiest Decembers in recent memory,[112] politicians, businessmen, beggars, adults, children, laborers, tourists, and neighbors gathered at the foot of the city's main wharf in the early dawn. The fog lifted to reveal a colorful and bustling crowd extending to the surrounding docks and hillsides. In the middle of the harbor lay *Colorado*, its bright white decks sparkling in contrast to its formidable, black wooden hull. Departing from the storied William H. Webb shipyards in New York, *Colorado* had circumnavigated the Americas, docking in Rio de Janeiro, Brazil, and Callao, Peru. Since the summer of 1865 it had worked the Pacific from San Francisco to Panama.[113] Today, thanks to a yearly subsidy of around half a million dollars for transporting mail,[114] it would embark on a massive pioneering enterprise for its giddy financers—a transpacific journey west to what was called "the Orient."[115]

After twenty-two days at sea, *Colorado* arrived safely at Yokohama. Once again, a cheerful crowd amassed to offer a remarkable welcome. "Whistles blew, cannons boomed, and the band on the French frigate *La Guerrière* played 'The Star-Spangled Banner.'"[116] Passengers disembarked, and cargo was unloaded throughout the night. The next morning, *Colorado* continued its trip towards Hong Kong. This time, it was not alone, as British and French steamers from the Peninsular & Oriental Steam Navigation Company (P&O) and the Messageries Maritimes regularly navigated the China Sea. It received a more discreet welcome upon arrival in Hong Kong, with "only" some forty-two gun-shots exchanged with the authorities on shore. On the night of January 30, 1867, *Colorado* finally completed its 9,700-kilometer journey from San Francisco to Hong Kong, successfully inaugurating PMSS's second line as well as the first regular passenger and cargo service across the north Pacific.[117]

---

112. Glen Conner, *History of Weather Observations San Francisco, California, 1844–1948* (Ashville, NC: Midwestern Climate Centre, 2005), 36, https://mrcc.purdue.edu/files/FORTS/histories/CA_San_Francisco_Conner.pdf.
113. "Launch," *New York Times*, May 22, 1864, 8; "The News," *New York Daily Herald*, September 23, 1865, 4; "Colorado," TheShipsList, accessed October 4, 2023, https://www.theshipslist.com/ships/descriptions/panamafleet.shtml.
114. Tate, *Transpacific Steam*, 24.
115. Oriental means situated to the east, but *Colorado* was sailing to a destination westward of the United States. Yet centuries of Eurocentric cartography made people from the United States still call China and Japan "the Orient."
116. Tate, *Transpacific Steam*, 25.
117. Tate, *Transpacific Steam*, 25–26.

Taking a closer look at *Colorado*'s first trip allows us to grasp the types of transoceanic passengers, some of whom will be the focus of subsequent chapters. In first class there was a small group of businessmen, headed by A. A. Low, the president of the New York Chamber of Commerce. There were also a few missionaries, militaries, and bureaucrats from the United States. By contrast, over 200 Chinese were traveling in steerage. Most of them were male laborers or small retail owners who were returning to China after years of work in the US on railroads, farms, stores, and mines. On the return voyage, started in Hong Kong on February 17 and ending in San Francisco on March 20, *Colorado* carried a similar distribution of passengers, making the Chinese the most numerous nationality on board.[118]

These hierarchies of class and nationality were also observed among the crew members, with the Chinese occupying a central role in transpacific voyages, just as they did during the era of the Manila galleons. On the first *Colorado* trip, officers and crew were predominantly from the United States, with only six Chinese who worked as steerage cooks and stewards to look after the Chinese passengers.[119] But soon thereafter, Allen McLane, PMSS's president, after traveling to China later in 1867 and realizing that European steamers all carried Chinese crews, made changes, as described in his report to the Board of Directors: "While in China I made a very important change in the personnel of our steamers, by the substitution in all departments, except as officers alone, of Chinese for European and American crews. The savings therefrom, in wages, food, etc., will be very great."[120] And they certainly were—Chinese were paid half what US stewards earned. Already by the end of 1867, *Great Republic*, the last of the PMSS steamers crossing the ocean that year, had 108 Chinese workers, including firemen, coal passers, servers, and steerage cooks, described later in a travel memoir as "the famous blue-gown boys. . . . Dressed in black silk caps, dark blue tunics and wide trousers, with their pigtails hanging down almost to the heels of their white-soled black felt slippers, they moved soundless over the zebra-wood floors with trays of champagne and cakes, serving each guest with a respectful bow and tiny slant-eyed smile."[121]

With the beginning of regular transpacific runs inaugurated by *Colorado*, San Francisco and the PMSS steamers consolidated as intermediaries for the traffic of people and merchandise between the Asian and Mexican coasts.[122] In the following

---

118. Tate, *Transpacific Steam*, 24–27.
119. Tate, *Transpacific Steam*, 240.
120. Cited in Tate, *Transpacific Steam*, 240.
121. Cited in Tate, *Transpacific Steam*, 240.
122. Sailing vessels continued to do the run, but in declining numbers. For information on the sailing vessels doing the San Francisco–Hong Kong run in the years after the inauguration of PMSS, see Sinn, *Pacific Crossing*, 124–33.

decades, the history of exchanges between the two regions would involve a series of attempts, some more successful than others, by Mexican, Japanese, and Chinese diplomats, bureaucrats, businessmen, sailors, and passengers to evade the monopoly of San Francisco and the PMSS. This is the story that the following chapters will explore.

2
Vasco de Gama, 1874
A Space Odyssey[1]

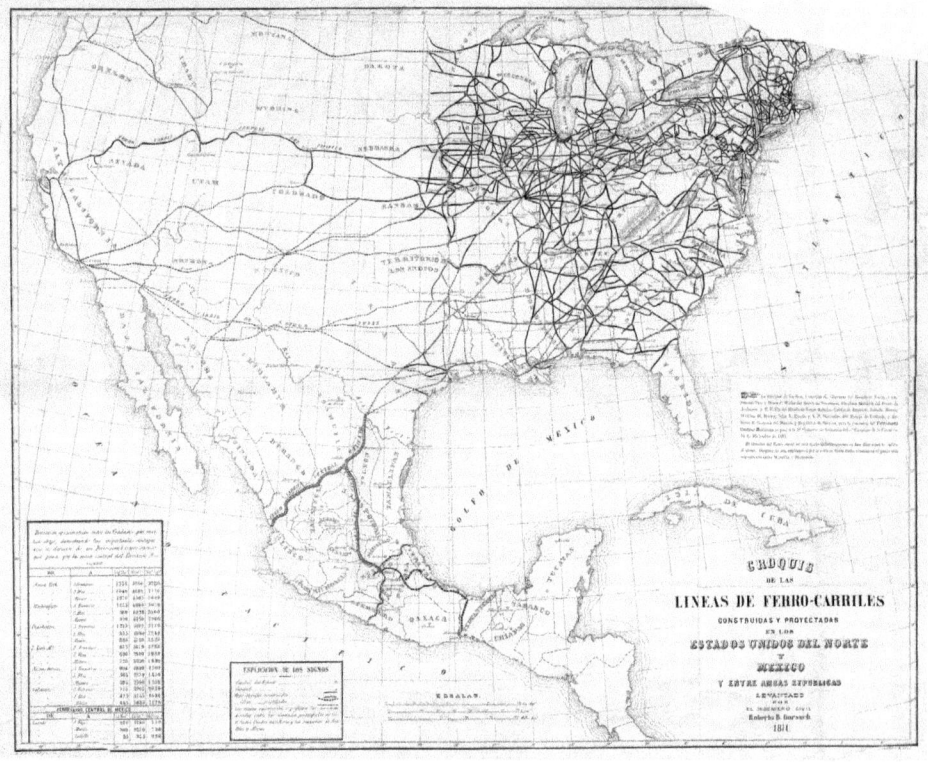

Map 2
Robert B. Gorsuch, *Líneas de ferrocarriles en los Estados Unidos del Norte y México* [Railway lines in the United States and Mexico]. Map. Mexico City: V. de Murguía e hijos, 1871. From Mapoteca "Manuel Orozco y Berra," Internacionales 2. Accessed December 15, 2023. https://mapoteca.siap.gob.mx/coyb-int-m50-v2-0055/.

---

1. The title was inspired by Marco Arturo Moreno Corral's book on the subject: *Odisea 1874 o El primer viaje internacional de científicos mexicanos* (Mexico City: FCE, 2003) and by Stanley Kubrick's 1968 science fiction film *2001: A Space Odyssey*.

An enormous wave crashed over the deck of the listing *Vasco de Gama*. It had been nineteen days since its San Francisco launch, and its pumps were screaming at full blast, bailing the onslaught of saltwater. Deep in the boat's bowels sat a fretting Francisco Díaz Covarrubias. With his large stature, manicured moustache, and three-piece suits, he usually came across as a self-assured, formal, and politically well-connected gentleman. In this late fall storm in the heart of the Pacific, he was reduced to the darkness of his thoughts in a bobbing berth on the high seas. Scribbling in his journal, he wrote, "the clouds obscured the sun's rise, making it impossible to determine the vessel's position in the deserted Ocean . . . The rain poured heavily and enormous storm waves swept across the bridge, flooding the entire deck. There weren't enough pumps to bail the water swamping the 'Vasco,' and the clamour produced [by the pumps] . . . added to the pitch of the roaring winds in such a way that not even by shouting was it possible to make oneself heard."[2]

Even worse for him than not being heard would be not being able to see. And that was already becoming a likelihood not just for him, but for the entire five-man scientific expedition that he was leading halfway across the globe. Díaz Covarrubias had hand-picked a team to chart the transit of Venus between the Earth and the Sun by presidential order. This was a once-in-a-century opportunity, not just to compete with other nations to measure the rare astral event, but perhaps more crucially to go down in the annals of history as the first successful official scientific mission sent abroad by a Mexican government.[3] They just needed to reach shore on time. Their plan "A" had already been rubbished—there was no way they would reach their original destination in Peking, as it was then known.[4] At this point just about anywhere would have to do—as long as they could set up their instruments on *terra firma*. Yokohama was the ship's first port of call.

Six weeks earlier, after years of lobbying, President Sebastián Lerdo de Tejada had finally authorized funds to set up an astronomic station in the so-called Far East, the area where the astral event was calculated to be most visible. By sending a Mexican Astronomic Commission, the country would join an exclusive group of nations that could afford to send costly delegations to observe the rare celestial phenomenon. The Europeans and the United States had planned their trips for years and had settled into their respective stations months in advance; the Mexicans, battered by decades of civil wars and foreign invasions, had left everything to the last minute.

---

2. Francisco Díaz Covarrubias, *Viaje de la Comisión Astronómica Mexicana al Japón para observar el tránsito del planeta Venus por el disco del sol el 8 de diciembre de 1874* (Mexico City: Políglota, 1876), 104. Translation by the author. The picture showing how Díaz Covarrubias as well as the other four traveling scientists looked comes from Moreno's book cover as well as Moreno, *Odisea*, 21.
3. Moreno, *Odisea*, 8.
4. Díaz, *Viaje*, 89.

As they sat on the open sea, Díaz Covarrubias could not help but regret that their mission, beleaguered by time constraints, lack of funds, and now foul weather, had become a living metaphor of the role played by Mexico in the concert of nations. Furthermore, their failure, if they did not reach shore on time, would discredit the country abroad and at home, where the untimely and costly expedition had been highly criticized by the opposition.[5]

Of the epic trip, two memoirs exist. Francisco Bulnes, the team's youngest member and its official chronicler, dedicated his 1875 volume to his benefactor, President Lerdo. A year later Francisco Díaz Covarrubias published his, with a special prologue for the minister of justice and public instruction.[6] The analysis and contextualization of this pair of documents serves as a means of exploring the ways in which the interrelated notions of imperialism, nation, progress, science, class, and race framed the experiences of transpacific travelers at the end of the nineteenth century. In particular, the case exposes how a group of educated, wealthy, and government-connected Mexicans viewed themselves and their relationship to Asian and other peoples within the asymmetries of the transpacific world and how their racist views, supported by allegedly scientific principles and an equally prejudiced press, set the tone to discriminate against Asians, notably Chinese, and limit their possibilities when traveling to and/or establishing themselves in Mexico. The contextualization of the memoirs also serves to determine the land and maritime routes and the ports that formed the network that linked the Mexican territory with the rest of the northern Pacific Rim, just before steam-powered technologies broadened their presence in the country.

Vasco de Gama and Transpacific Steam Passenger Services in the 1870s

The Mexican Astronomic Commission's journey took place in a world of nations in the making. While this could be said for many historical periods, it was particularly evident for the Americas in the nineteenth century, when all the Pacific nations with the exception of the United States obtained political independence: Mexico, along with its southern neighbors, emancipated from Spain in the 1820s; the Canadian Confederation did so from England in 1867, with British Columbia joining it in 1871. Even the United States, whose war of independence preceded

---

5. Moreno, *Odisea*, 16–17, 22–24.
6. Francisco Bulnes, *Sobre el Hemisferio Norte, once mil leguas. Impresiones de viaje a Cuba, los Estados Unidos, el Japón, China, Conchinchina, Egipto y Europa* (Mexico City: Revista Universal, 1875); Díaz, *Viaje*. The author consulted the original editions available online on Mexicana, https://mexicana.cultura.gob.mx/, an open platform that offers the digitalized collections of the Mexican Ministry of Culture. The author and Martin McLennan translated quotations from these works into English. There was a third publication that came out of the trip, *Observaciones del tránsito de Venus hechas en Japón por la Comisión Astronómica Mexicana*, printed in Paris in 1875, covering only the technical information of the astronomic measurements. Moreno, *Odisea*, 8, 119.

all others, did not consolidate its presence in the Pacific, at the expense of Mexico, until the mid-century. Furthermore, national borders generally began to be enforced during the lifespan of Díaz Covarrubias and his team. In fact, two of the Commission's members, Francisco Jiménez and Manuel Fernández Leal, participated in the geographic demarcation of the Mexico–US border between 1849 and 1856.[7]

Empire-making directly framed the Commission's experiences. For instance, as discussed in the previous chapter, the three Pacific coastal cities where *Vasco de Gama* and its Mexican passengers docked—San Francisco, Yokohama, and Hong Kong—had rapidly and recently transformed from small towns into large international ports because of their strategic value for the United States, Japan, and England respectively.[8] The monopoly of transpacific passenger services and steam technologies was also related to imperial interests. By the 1870s, only the United States and Great Britain had been capable of subsidizing regular transpacific services and a steam merchant fleet, as evidenced by the fact that all the steamships boarded by the Mexican Astronomic Commission on their way to Asia, with the exception of a small French vessel in the Atlantic portion of the trip, had been built either in US or British shipyards. Chapter 1 already explained the case of the Pacific Mail Steamship Company (PMSS), the first to establish regular transpacific passenger services. PMSS's first real competitor was the China Trans-Pacific Steamship Company (CTPC), created in June 1872 by a group of British merchants living in Hong Kong who received imperial subsidies for the venture. It started operations a year later with the trial run of *Galley of Lorne*, which carried 641 Chinese passengers in steerage and set a new record by crossing from Hong Kong to San Francisco in just twenty-six days. Subsequently, regular operations began with two brand-new British-manufactured steamers, *Vasco de Gama* and *Vancouver*. The company advertised *Vasco* as "built especially for the Oriental trade, fitted up most luxuriously with all the latest improvements for the comfort and safety of passengers."[9] The CTPC was short-lived as it could not compete with a new venture established in San Francisco by the transcontinental railroad owners—the Oriental and Occidental—nor with PMSS's reduced passenger fares.[10] As a consequence, by

---

7. Moreno, *Odisea*, 36–40.
8. As discussed in Chapter 1, barely three decades before the Commission's voyage, San Francisco had passed from Mexican to US hands. This, combined with the discovery of gold a few months later, triggered a sudden growth that catapulted San Francisco into a bastion for exploration and incorporation of the surrounding areas for the US. It also turned it into the launching pad for their imperial project in the Pacific. In the case of Hong Kong, the British occupation since 1842 triggered its sudden expansion as it became a base for their imperial interests in the region. See Chapter 1 for more information and sources. In the case of Yokohama, it had been barely a decade since the Japanese government had opened up to foreign trade, forced by a combination of local reforms and foreign pressures. The latter will be further discussed in Chapter 4.
9. Cited in E. Mowbray Tate, *Transpacific Steam: The Story of Steam Navigation from the Pacific Coast of North America to the Far East and the Antipodes, 1867–1941* (New York: Cornwall Books, 1986), 29.
10. See Chapter 4 for a brief description of the Oriental & Occidental Company.

the time the Mexican scientists boarded *Vasco*, it had already been chartered and operated by PMSS.[11] As such, its high officers were from the United States and many had previously worked for the navy.[12] Its crews were Chinese.[13]

Within this imperial mapping, and considering the political and economic instability that had characterized Mexico since its independence, the country relied on foreign capital and a policy of subventions to incorporate steam technologies for transportation. In the case of railroads, the first line covering the 660 kilometers from Mexico City to Veracruz on the Atlantic coast, which was to be used by the Mexican scientists, was completed in 1873 and was owned by British businessmen.[14] As for steamers, the first had arrived in the Gulf of Mexico at the beginning of the 1840s,[15] then on the Pacific coast less than a decade later.[16] All were foreign vessels that moved around without Mexican intervention, so their itineraries depended on private interests. In order to regulate the sector and guarantee more regular sailings, the Mexican government started a policy of subventions to these foreign companies in 1867.[17] The first was granted to Alexander and Sons for a route linking New York with Veracruz and Sisal, in Yucatán.[18] In the Pacific, the first companies to be subsidized were the PMSS and the Panama Railroad Co. in 1872, followed by the Pacific Steamship Co. in 1875, the California and Mexico Steamship Co., and the Accelerated of the Gulf of Cortes in 1877. All were owned by US investors and circulated along the San Francisco–Panama route.[19]

---

11. Tate, *Transpacific Steam*, 29; Elisabeth Sinn, *Pacific Crossing: California Gold, Chinese Migration, and the Making of Hong Kong* (Hong Kong: Hong Kong University Press, 2013), 131–32.
12. Tate, *Transpacific Steam*, 238.
13. Chinese crews worked on PMSS's transpacific routes until the passage of the Seamen's Act in 1915. It stated that 75 percent of the crew of any ship over one hundred tons had to understand commands given in English. "The intent was to eliminate Chinese crews. However, the U.S. Department of Commerce ruled that such practices as sign language or the use of pidgin English could meet the requirement, and many crews were continued on this basis." Tate, *Transpacific Steam*, 241.
14. Priscilla Connolly, *El contratista de Don Porfirio. Obras públicas, deuda y desarrollo desigual* (Mexico: ColMich/UAM/FCE, 1997), 82; Sandra Kuntz Ficker, "Fuentes para el estudio de los ferrocarriles durante el Porfiriato," *América Latina en la historia económica. Boletín de fuentes* 13–14 (January–December 2000): 137.
15. Karina Busto Ibarra, *El Pacífico mexicano y sus transformaciones: integración marítima y terrestre en la configuración de un espacio internacional, 1848–1927* (Mexico City: El Colegio de México, 2022), 57. See also Mario Trujillo Bolio, *El Golfo de México en la centurión decimonónica. Entornos geográficos, formación portuaria y configuración marítima* (Mexico City: Porrúa/CIESAS/Cámara de Diputados LIX Legislatura, 2005).
16. As mentioned in earlier chapters, PMSS vessels were the first to touch Mexican ports regularly on their runs from Panama to San Francisco, but there were also those from the Nicaragua Line, which docked at Mazatlan and Acapulco, and those from the Colorado Steam Navigation Company, stopping in various places on the Gulf of California. Busto, *Pacífico mexicano*, 57–58.
17. Busto, *Pacífico mexicano*, 58.
18. The contract stipulated that the company would receive 2,000 pesos per trip as well as up to 7,000 pesos of tax exemptions per year as long as its vessels made at least eighteen trips per year to Veracruz and Sisal. Roberto García Benavides, *Hitos de las comunicaciones y los transportes en la historia de México* (Mexico City: Secretaría de Comunicaciones y Transportes, 1988), 175.
19. According to the contract, PMSS would receive 2,500 pesos per month and several tax exemptions if its steamers stopped in Acapulco, Manzanillo, Mazatlan, and San Blas (Mexican Line) and in San Benito, Tonalá, Salina Cruz, and Puerto Ángel (Central American Line). Busto, *Pacífico mexicano*, 66–67. The Panama Railroad Co., on its part, would receive 27,500 pesos per year for twelve trips covering Mexican southern ports, as well as

Troubled Politics, the Rise of Positivist Science, and the Astronomic Commission's Conception

The first diplomatic contacts between Mexico and an Asian country occurred not long after Napoleon III's 1862 successful invasion of Mexico.[20] Soon after assuming power in 1864, Maximilian of Hapsburg requested D. G. Overbeck, then Austrian consul in Hong Kong, to act as the representative for Mexico in China. Despite his immediate acceptance, the British authorities delayed the appointment due to the absence of a treaty between Mexico and China. By the time official approval arrived in the summer of 1866, Overbeck was departing for Shanghai and left his credentials for Adalf Eimbcke, who was both a personal friend with previous Mexican experience and a partner at the local firm Caslovitz & Co.[21] However, by this time Maximilian's government had been severely weakened due to the gradual repatriation of French troops, the loss of many of his local supporters, and the increasingly successful attacks of the republican armies led by Benito Juárez, who considered himself the legitimate president of Mexico. In this context Eimbcke's appointment did not materialize. Nonetheless, a pioneer group of Cantonese laborers arrived in the northern state of Chihuahua to work on the construction of the Central National Railroad. They were hired by the newly established Chinese Colonization Company, which planned to bring in laborers over the course of ten years but most likely disbanded beforehand.[22] Another group arrived in Veracruz to work on the plantations of Manuel B. da Cunha Reis.[23] In the following years, a few other contingents arrived from California to work in Baja California's gold mines and cotton

---

various tax exemptions. Nevertheless, this contract was never put into effect. García Benavides, *Hitos de las comunicaciones*, 176. See a list of all the maritime companies that received subventions from the Mexican government on both coasts between 1867 and 1920 in Busto, *Pacífico mexicano*, 62–65.

20. This book uses the name "Mexico" for the independent country. When referring to the colony of Spain between 1521 and 1821, we use the name given to it by the crown, "New Spain." The only antecedent of Asian presence in Mexican diplomatic papers known up to now comes from the Foreign Affairs Commission formed by Juan Francisco Azcárate y Ledesma, Manuel de Heras y Soto, and José Sánchez Enciso immediately after Mexico obtained its independence from Spain. In December 1821, they released a document that addressed the importance of having relations with Asian countries in order to sell products, make ships in Cavite, Philippines (which had been New Spain's main shipyard), export silver coins, and bring Asian people, notably Chinese, to colonize Texas and Alta California, the provinces that Mexico eventually lost to the United States in the 1846–1848 war described in Chapter 1. But this did not materialize into diplomatic contacts until 1864. Mercedes de Vega et al., *Historia de las relaciones internacionales de México, 1821–2010*, Vol. 6 (Mexico City: SRE, 2011), 58–63.
21. Archivo General de la Nación de México (AGNM), Administración Pública Federal S. XIX/Gobernación S. XIX, Relaciones Exteriores, caja 94, expediente 73, 1864–1866, "Expediente sobre el nombramiento de G. Oberbeck como cónsul del Imperio en HK," 15–16.
22. José Jorge Gómez Izquierdo, *El movimiento antichino en México (1871–1934): problemas del racismo y del nacionalismo durante la Revolución Mexicana* (Mexico City: INAH, 1991), 56; Raymond B. Craib III, *Chinese Immigrants in Porfirian Mexico: A Preliminary Study of Settlement, Economic Activity and Anti-Chinese Sentiment*, Latin American Institute, University of New Mexico, Research Paper Series 28 (Albuquerque: University of New Mexico, 1996), 6.
23. Elliott Young, *Alien Nation: Chinese Migration in the Americas from the Coolie Era through World War II* (Chapel Hill: University of North Carolina Press, 2014), 107.

plantations and to continue working on the area's railroads.²⁴ Maximilian's government ended abruptly with his execution by the republican armies in 1867. Since his rule was the product of a foreign intervention and ran parallel to the liberal government of Benito Juárez, this episode of Mexico–Asia relations has received little official acknowledgment.²⁵

In 1867, the same year that the PMSS started its transpacific runs as described in Chapter 1, the heterogeneous group of liberals that supported President Juárez arrived triumphantly in Mexico City. They finally began to enjoy a period of relative peace after fighting and defeating the conservative armies in two consecutive conflicts:²⁶ a civil war that lasted from 1858 to 1861, followed by a conservative-backed French invasion between 1862 and 1867. At the heart of both wars had been the conservative opposition to the Constitution of 1857 and the so-called Reform Laws that preceded it. In general terms, these legal documents promoted the secularization of society, the defense of individual over collective rights, a republican form of government with regular elections, and equality under the law for all citizens, at least on paper. These measures had upset the legal, political, and economic privileges of traditional corporate entities such as the army, guilds, Indigenous communities, and particularly the Catholic Church. While all these institutions had the support of ample sectors of the population, the pact made by some of their leaders with a foreign monarch, Maximilian of Habsburg, and their subsequent military defeat eroded their strength and legitimacy and propelled the success of Juárez's group.²⁷ Juárez died in 1872 and was succeeded by his next in rank, Sebastián Lerdo de Tejada, who later won the elections for the period 1872–1876.²⁸

One of the first measures implemented by Juárez's government, of which the scientist Díaz Covarrubias was part, was a profound educational reform meant to secularize and, in the words of those who headed it, "regenerate" Mexican society.

---

24. Shicheng Xu, "Algunas reflexiones sobre el desarrollo de las relaciones sino-mexicanas," *Cuadernos Americanos* 121 (July–September 2007): 171–86; Gómez, *Movimiento antichino*, 56; Craib, "Chinese Immigrants," 6.
25. The exception is De Vega, *Historia de las relaciones*, 65–66. For more on Maximilian's government and the French intervention, see Konrad Ratz, *Tras las huellas de un desconocido: nuevos datos y aspectos de Maximiliano de Habsburgo* (Mexico City: Conaculta/Siglo XXI, 2008); Konrad Ratz, *Maximiliano de Habsburgo* (Mexico City: Planeta, 2002).
26. After 1867 the major military threat came not from the conservatives but rather from within the liberal ranks, when General Porfirio Díaz challenged the results of the presidential elections of 1871 and 1876. Díaz's prominence in Mexican politics and the repercussions of his governments for transpacific relations will be addressed in the upcoming chapters.
27. On the subjects of Mexican Liberalism and the different groups and ideologies that formed part of it as well as the Reform Laws and the Civil War, see Charles Hale, *The Transformation of Liberalism in Late Nineteenth-Century Mexico* (Princeton, NJ: Princeton University Press, 1989); Josefina Zoraida Vázquez, ed., *Interpretaciones del periodo de Reforma y Segundo Imperio* (Mexico City: Patria, 2007); Colin M. MacLachlan and William H. Beezley, *Mexico's Crucial Century, 1810–1910: An Introduction* (Lincoln: University of Nebraska Press, 2010), 75–103.
28. Sebastián Lerdo de Tejada was the head of the Supreme Court when Juárez died. As such, he assumed the interim presidency and later won the elections against Porfirio Díaz, whom we will talk about in the subsequent chapters. Juárez was president of the Republic from 1858 until his death in 1872. For an overview of his biography and governments, see Brian Hamnett, *Juárez* (New York: Longman, 1994).

The main product of this reform was the law of December 2, 1867, that created the Escuela Nacional Preparatoria (ENP, National Preparatory School), that is, a five-year uniform preparatory curriculum that followed primary studies and would prepare students for a professional education. The commission in charge of evaluating and implementing the changes was formed by a group of scientists who held administrative posts in the Juárez administration and were also linked to each other by family ties. The most prominent of them were its president Gabino Barreda, Pedro Contreras Elizalde, and the brothers José and Francisco Díaz Covarrubias—the latter would become the head of the Mexican Astronomic Commission that traveled to Asia in 1874.[29] Both Contreras and Barreda had studied medicine in Paris, where they had met and been influenced by the positivist philosopher Auguste Comte, who argued that man's only way of knowing was through the scientific method, composed first by observation and experiment, and second by a search for the laws of phenomena and their interrelations. Science, as they understood it, would inevitably make humanity progress and reach a positive era where scientific thought would prevail over supposedly inferior forms such as theology and metaphysics, which they linked to earlier stages of mental development. As a consequence, the ENP had a positivist curriculum with the sciences at its core, supplemented by the learning of languages—mainly so that students could understand scholarly texts in their original languages—and a few humanities.[30] Besides teaching a new generation of students the proper methods of science, it was expected that this intellectual education would enable future leaders to lead the reconstruction of society by spreading knowledge among the so-called ignorant masses. Since independent thought outside the curriculum was seen not only as fundamentally wrong but also as an obstacle to progress, the ENP's adherents defended these principles dogmatically.[31]

In this context, where global elites influenced by positivism equated desirable progress with scientific and technological advancements, observing the transit of Venus was no mere intellectual hobby. Rather, it was the only means to obtain an essential measurement for the international scientific community: the Astronomic Unit, that is, the distance between the Earth and the Sun. Since Nicholas Copernicus published *De Revolutionibus Orbium Coelestium* in 1543, making the Sun and not the Earth the center of the solar system, scientists had been trying, unsuccessfully, to measure the distance between the Sun and its planets. In 1716 British astronomer Edmond Halley forwarded a possible solution. In *A New Method of Determining the*

---

29. Pedro Contreras Elizalde was married to one of Juárez's daughters; Gabino Barreda married José's and Francisco's sister. Hale, *Transformation of Liberalism*, 141.
30. The ENP's initial curriculum was as follows: in the first and second years, students took mathematics, beginning with arithmetic and ending with calculus, as well as English and French. In the third year they took mechanics, elementary astronomy (called cosmography), Greek, and Spanish grammar. The last two years included the following courses: chemistry, natural history (which comprised botany and zoology), logic, Latin, geography, history, and literature. Hale, *Transformation of Liberalism*, 143.
31. See Hale, *Transformation of Liberalism*, 139–68.

Parallax of the Sun, or His Distance from the Earth, he suggested that the Astronomic Unit could be obtained by measuring the beginning and the end of the transit of Venus across the Sun from at least two different places on Earth. However easy it sounds, his method was difficult to test, as Venus orbits between the Earth and the Sun in subsequent intervals of eight and 122 years, giving a generation of scientists only two chances to succeed. Astronomers put Halley's method into practice after his death, in 1761 and 1769, but they never achieved consensus on their measurements. The subsequent transit, in December 1874, represented a renewed opportunity to test Halley's method, one that might lead to the ability to determine the correct dimensions of the entire solar system.[32]

With this in mind, several countries had prepared their astronomic commissions well in advance. For instance, in 1872, the French had destined some 400,000 francs to install six stations in Asia. The United States, on its part, spent over $200,000 and took three years to organize its eight commissions. Germany sent five groups to Asia and Africa, and Russia set up twenty-five commissions throughout its Asian territory. The British crown financed five groups while Lord Lindsey funded his own private expedition. And finally, a group of Italians installed an observation post in northern India.[33] In Mexico, the Chamber of Deputies had commenced a tepid discussion on the subject in 1871. But as the government's finances had not recovered from the war against the French, concluded only four years earlier, the project was soon abandoned. Mexican scientists, on their part, had tried to raise public awareness of the importance of sending a commission through the publication of academic and informational articles. They argued that the mission was relevant not only for its possible contribution to science and humankind, but also for generating international prestige for the country.[34]

Their attempts remained unnoticed until three months before the deadline. On September 8, 1874, during the celebrations of the twenty-seventh anniversary of the defense against the US invasion of Mexico, deputy Juan José Baz brought the topic to the attention of President Lerdo de Tejada.[35] Days later Lerdo conferred with Francisco Díaz Covarrubias, one of the country's most renowned astronomers, who was also an ENP professor.[36] While the latter expressed confidence in local scientists' capacity to perform, he was concerned about the time available to mount the expedition and travel to Peking, his ideal point of observation. According to his calculations, it would take at least forty-five days to travel from Mexico City to

---

32. Moreno, *Odisea*, 11–16.
33. There is a list of the specific places where these commissions settled in Bulnes, *Sobre el hemisferio norte*, 160.
34. Moreno, *Odisea*, 17–21.
35. Hugo Diego, *Viaje al Japón. Francisco Díaz Covarrubias* (Mexico City: Ediciones de Educación y Cultura, 2008), 12.
36. Besides participating in the commission that created the ENP and being one of their professors, Díaz Covarrubias held different posts during the Juarez governments, including director of the National Astronomic Observatory when the French troops invaded Mexico. He refused to work for them and remained loyal to Juárez until his death. See his biography in Moreno, *Odisea*, 27–35.

Yokohama, plus an additional ten days to get to Peking, arriving just days before the event. On September 14, Díaz Covarrubias and President Lerdo agreed that a five-member astronomic commission would be sent to Asia. This number would allow for the group to split up and make two separate measurements to test Halley's method without depending on foreign results. Díaz Covarrubias chose three of the country's finest and most experienced engineers to accompany him: Francisco Jiménez, Manuel Fernández Leal, and Agustín Barroso. All had performed astronomic observations in the past and the latter was also a distinguished photographer. President Lerdo added one last member, Francisco Bulnes, a young engineer and talented writer who would be in charge of recording the memoirs of the trip. Besides their governmental posts and personal scientific projects, all but Francisco Jiménez taught mathematics in the ENP by the time they left for Asia.[37]

An Astronomic Transpacific Journey

The closest port to Mexico City with direct passenger services to Japan and China was San Francisco. To get there, there were two main options. The shortest was to travel by land to the nearest western port, Acapulco or Manzanillo, and then take a steamer to San Francisco. As there were no railroads linking Mexico City with the Pacific coast and the dirt roads remained in terrible shape from the rainy season, this option was discarded. Díaz Covarrubias thus opted for a longer but safer route. He would take the only existing railroad in the country—inaugurated by President Lerdo only a year earlier—towards the port of Veracruz on the Atlantic coast, sail to New York, then take the transcontinental railroad across the United States all the way to San Francisco. According to Díaz Covarrubias's calculations, this first portion of the trip would take around twenty days. On the night of September 18, 1874, after rapidly gathering and carefully packing all the delicate equipment, the scientists left Mexico City with the hope of reaching Asia on time to do the observation.[38]

The first obstacle for the mission came in the form of viruses. The first destination, the port of Veracruz, was experiencing an epidemic of yellow fever. If any of the Commission's members fell sick, their success would be compromised. Díaz Covarrubias thus decided to wait for the steamer in the highlands, halfway between Mexico City and Veracruz. On September 22, he finally received a telegram stating that *Caravelle*, a small, 800-ton French steamer, had docked. Unfortunately for him, *Caravelle* was leaving two days later for Havana, not the United States. But with so many days already wasted, Díaz Covarrubias decided to go to Cuba, where he had a better chance of catching a boat for New York. *Caravelle* took the commissioners

---

37. Díaz Covarrubias also served as the ENP's assistant director. Francisco Jiménez also taught math but, due to his military formation, he did so in the army's school, called *Colegio Militar*. Moreno, *Odisea*, 34–43.
38. Bulnes, *Sobre el hemisferio norte*, 13. According to Moreno, *Odisea*, 23–26, Francisco Jiménez and Manuel Fernández left the following day and quickly caught up with the initial group formed by Díaz Covarrubias, Barroso, and Bulnes.

from Veracruz to Havana in less than five days at an average speed of thirteen kilometers per hour. Díaz Covarrubias described it as "really modest: its small size made it intolerable to stay on deck due to the proximity of the heat coming from the chimney; and the kitchen, located between-decks and towards the bow, did not allow us to ignore the meals' preparation nor the multiple smells, especially repugnant at sea. Nevertheless, most passengers, even if seasick, remained atop, as the temperature inside the narrow chambers was [even more] unbearable."[39] The heat made it impossible for Díaz Covarrubias to "do anything productive but to read at times, [so] this first part of the journey was truly tedious as the *Caravelle* had no piano, no chess, nor any other means of distraction" that characterized the first-class services of larger steamers.[40] Bulnes, with his more literary tone, described his first days at sea as follows: "The first days of navigation are sad. The passengers in the first period of seasickness guard their silence and go horribly pallid; in the second, they empty their stomachs; in the third, they remain dazed like the lotus eaters. The women forget their place among men and fall quickly to human misery, losing their angels' wings in the convulsions of the sickness."[41] On September 28, the vessel reached Havana, where the Mexicans spent two days waiting for the next steamship to depart. Díaz Covarrubias found the city "notably decayed in comparison to when I was first here fourteen years ago. The effects of the civil war devouring this wealthy [Spanish] colony are now visible, . . . [although] it still remains full of movement and active trade [with] multiple vessels from all European and American maritime nations crowding the large piers."[42] On September 30, after stamping their passports with the Cuban delegate and the Mexican consul, the five Mexicans, along with a German, two Cubans, and one passenger from the United States, set sail for Philadelphia in the 1,400-ton US steamer *Yazoo*.[43] The drop in temperature, the ample space, and the tranquil waters made this a more pleasant journey. Bulnes wrote: "Between Havana and Philadelphia we covered a distance of 1,100 miles without the slightest delay. Having left the cape of the Antilles, everything seemed better. We were joined by a journalist from New Orleans, and a Prussian officer that preferred Sedan over Cuba's Obispo street."[44] On October 5, as everyone was getting ready to disembark, a sanitary officer came aboard and announced quarantine for the vessel, as it was coming from plague-infested ports. Already behind schedule, the Mexican commissioners could not afford this additional delay. Captain Barret, the only one allowed off board, offered to contact the Mexican consulate and to

---

39. Díaz, *Viaje*, 48–49.
40. Díaz, *Viaje*, 50.
41. Bulnes, *Sobre el hemisferio norte*, 19.
42. Díaz, *Viaje*, 51–52.
43. Díaz, *Viaje*, 57. The steamer's name was spelled *Yat-soo* in Bulnes, *Sobre el hemisferio norte*, 35.
44. Bulnes, *Sobre el hemisferio norte*, 35–36. Bulnes's pages are full of historical and literary references, often with a sarcastic twist. Sedan was the battle where Napoleon III, who had ordered the invasion of Mexico in 1862, was captured by the Prussian army. The Obispo street is one of the oldest and most central in Havana, existing up to the present.

persuade the port authorities to revoke the order, since no passengers presented any symptoms. After sunset, the captain returned with good news for the Mexicans: They were allowed to leave.[45] This is reminiscent of the preferential treatment given to first- and second-class passengers in the case of the steamer *Suisang*, briefly described in the Introduction, and contrasted with the forced quarantine that the steerage passengers had to endure, regardless of their actual state of health.[46]

On October 6, the troop took the train for New York where, after a day of wandering around and getting supplies, they boarded the transcontinental railroad, which took seven days to cross the 5,300 kilometers separating New York from San Francisco. "The first thing that called my attention," wrote Bulnes, "was the luxury . . . of the Pullman palace sleeping car. . . . Every passenger has his own sofa and it's useless to look out of the windows, because an infinity of mirrors introduce their own varied and splendid landscapes in the interior where the comfort of a civilization which responds to the fantasies of the strangest of dreams is elegantly situated."[47] Díaz Covarrubias also remarked on the "many comforts" that made this train "superior to everything we saw in Europe."[48] He found the rapidity with which the salon transformed into a dormitory fascinating. The speedy train ran at thirty-six kilometers per hour and made regular meal stops, of which Díaz Covarrubias was highly critical: waiters were few and rude and had very little time to serve a large and hungry crowd: "we were [thus] forced to eat really quickly, badly and very little or to incur, despite our disgust, into the Anglo-American habit of grabbing a piece of ham, cheese or bread and eat it while standing or sitting inside the cars."[49] The transcontinental journey came to an end on October 14, when the train reached Oakland and a ferry boat took all passengers and cargo across the bay to San Francisco.[50]

Díaz Covarrubias was impressed with San Francisco. In his words, it was "insignificant thirty years ago [and] now has close to 180,000 inhabitants. Its magnificent bay, its increasing trade with Asia, . . . the gold discovered nearby, and the frenzy of activity of the Anglo-American race perfectly explain its rapid growth."[51] But all this made the city expensive: "In the purchase of a few small objects I had to invest a sum four or five times larger than the amount I would have invested in Mexico or even in New York."[52] The Mexican travelers suffered a new delay in San Francisco. The next steamship for Asia would not leave for another five days. Díaz Covarrubias had originally calculated twenty days to get to San Francisco. So far,

---

45. Díaz, *Viaje*, 61–62.
46. The *Suisang* case will be thoroughly analyzed in Chapter 5.
47. Bulnes, *Sobre el hemisferio norte*, 71–72.
48. Díaz, *Viaje*, 72.
49. Díaz, *Viaje*, 77.
50. Díaz, *Viaje*, 86–87.
51. Díaz, *Viaje*, 90.
52. Díaz, *Viaje*, 90.

they had spent twenty-seven. With this additional delay they were already twelve days behind schedule.

Díaz Covarrubias used the forced wait as a chance to reconsider the location of the observation. After interviewing the Japanese and Mexican consuls and the engineers at the Coast Survey, one of the agencies participating in the United States' astronomic commission, Díaz Covarrubias opted for Japan over China. He listed the reasons as follows. First, the war between the two countries for the possession of the Liu Chiu islands seemed imminent, and Díaz Covarrubias felt safer in Japan due to its stronger maritime and military capacities. The second reason was "the frank hospitality that the current enlightened government of Japan affords to foreigners" as opposed to the Chinese "intolerant and even hostile [ways] to everything that comes from abroad."[53] Third, the weather near Yokohama seemed favorable for Díaz Covarrubias's observations as clouds and storms were infrequent during winter. And finally, staying in Yokohama saved at least a week of travel so that the Commission could recover some lost time.[54]

On October 19, the Mexicans set sail for Yokohama aboard *Vasco de Gama*, a 3,000-ton British steamer operated by the PMSS after the CTPC's bankruptcy. Powered by both steam and sail, *Vasco* was able to travel fast, at an average of eighteen kilometers per hour, that is, half the speed of the transcontinental train, but a third faster than the small steamers that the troop had taken in the Atlantic. While this was encouraging for the Mexican commissioners, who had such tight time constraints, the winds made this a "painful journey [with] only two days [of] calm sea; in the remaining eighteen a strong Northern wind forced us to port sending us night and day the bitter water of the waves. . . . no one was able to eat, sleep, walk, or remain seated; life under this oscillating regime damaged all nutritional functions, and left a profoundly pallid stain on all faces. Never-ending massive lead-colored waves rose to the heights of ten and twelve metres."[55] The Mexicans shared the first-class stern compartments with some fifty other passengers. Most of them were English-speaking doctors who were establishing enterprises in China or Japan.[56] There were also four or five Europeans. Among them was a German hat trader who planned to settle in Japan, who soon after landing went into bankruptcy simply because locals did not wear hats. There was also a Belgian specialized in gun powder who saw in the Sino-Japanese confrontation a good opportunity for profit. About the handful of women travelers, Díaz Covarrubias simply commented that during the few episodes of calm sea, the male passengers amused themselves by walking on deck, "guiding the ladies by the arm," although frequently "a treacherous wave would interrupt the pleasant conversation, spilling over the talkers a deluge soaking

---

53. Díaz, *Viaje*, 88–89.
54. Díaz, *Viaje*, 87–89.
55. Bulnes, *Sobre el hemisferio norte*, 91–92.
56. Díaz, *Viaje*, 95.

them from tip to toe, and forcing them to run to their cabins to change clothes."[57] The piano and singing recitals also kept them busy—that is, until the pianist was launched to the floor by the ship's sudden encounter with large waves. And much more entertaining than listening to the choir sing was observing their awkward positions while trying to stand still: "more than musicians, they looked like fencers in a fight."[58] Peeking into the bow's steerage compartments became another form of entertainment for Díaz Covarrubias. He observed "countless Chinese . . . [who] piled into those cabins . . . [and] must have been extremely uncomfortable during the storms; but they never stop playing . . . that game with dice and some sort of dominoes similar to the European one. Even though many of them were coming back to their country to enjoy their earnings after having worked in America for years, their clothing was the same than those worn by the poorest of them."[59] On the night of November 7, after eighteen days of painful trajectory, *Vasco de Gama* reached the outskirts of Yokohama. Yet "the threats of the ocean were felt more than ever. . . . The only way to overcome the tenacity of the dense fog in the bow of ship, was to wait. For about an hour we bobbed about without barely advancing."[60] Some of the passengers suggested deviating to calmer southern waters to avoid entering Yokohama in the middle of a storm in the dark, risking a possible crash. But *Vasco* did not alter its course. "Around midnight we finally saw [the lighthouse of Cape Kii] peering through the mist on the horizon, and it was saluted with a general hooray! We were in Asia. . . . Before sunrise *Vasco* had dropped anchor in the bay of Yokohama, some 250 metres from the city's piers."[61]

Just like San Francisco, Yokohama had grown from a relatively small fishing village into a large international port in barely a decade. The scientists' memoirs described it as comprising two sections. The four to five thousand foreigners—mostly diplomats from Europe, the United States, and Peru—lived in luxurious mansions on a hill known as the Bluff. The Japanese lived to the north, in the sister city of Kanagawa. While the two sectors got along, the twenty-one shots fired by every military vessel arriving in Yokohama and the equal response given by the Japanese authorities made Bulnes sarcastically remark that "the relationship between the Asian and European civilizations gets activated by the cannon."[62] The Mexicans found the city noisy at night. Besides the gunshots, the tam tams of the police patrolling the city or the firemen responding to a call,[63] as well as the whistles

---

57. Díaz, *Viaje*, 97.
58. Díaz, *Viaje*, 97.
59. Díaz, *Viaje*, 97–98.
60. Bulnes, *Sobre el hemisferio norte*, 92.
61. Díaz, *Viaje*, 106.
62. Bulnes, *Sobre el hemisferio norte*, 96.
63. Lionel Frost argued that during the rapid urbanization of major Pacific ports during the second half of the nineteenth century, the cities experienced regular fires as buildings were made of wood and the rapidity of the construction was done at the expense of safety measures. See Lionel Frost, "Rim of Fire: Pacific Rim Cities and the Problem of Fires," in *Studies in Pacific History: Economics, Politics, and Migration*, ed. Dennis O. Flynn, Arturo Giráldez, and James Sobredo (Burlington, VT: Ashgate, 2002), 108–22.

of the blind men who wandered the streets delivering massages and relaying the news of the day, woke them regularly. With the mediation of Mr. Bingham, the US ambassador to Japan, Díaz Covarrubias was able to meet with Mr. Terashima, the emperor's prime minister. He rapidly granted the necessary permits to establish an observatory in each of the city's two districts, emphasizing that "the emperor is decided to take in the emissaries of science as they deserve it."[64] The Mexicans hired a Chinese carpenter, Mow-Cheong, to build the wooden structures for the two observatories. Díaz Covarrubias described him as "hardworking and quite clever," performing his job with "care and thoroughness that are entirely Chinese."[65] By November 30, Mow-Cheong had finished the observatories and a Mexican flag was raised in both. The Commission was ready for the transit of Venus.[66]

The weather, though, was beyond their control. Despite the references given by locals concerning Yokohama's fair climate, the day before the event the city remained overcast. One hundred and five years earlier, the French astronomer Le Gentil had faced a similar obstacle. After spending all his money and reputation to get to Pondicherry, India, in time for the 1769 transit, he could not make any measurements because the clouds blocked the entire event.[67] Fearing a similar fate, none of the five engineers dared to speak a word during the last meal before the transit. They were fortunate enough to wake up to a clear sky on the morning of December 9. Just past 11 a.m., Venus became visible in Díaz Covarrubias's observatory. Minutes later, in the opposite part of town, Francisco Jiménez and Manuel Fernández saw it, too. The event lasted close to four hours. Both teams were successful in their measurements and thirteen out of the seventeen pictures taken by Agustín Barroso came through.[68] Their success was rapidly transmitted to President Lerdo with the following telegram: "To D. Sebastian Lerdo de Tejada, President of the Mexican Republic. Mexico. Complete success in the observations. Please receive my most sincere congratulations. F. Díaz Covarrubias C. Yokohama, Dec. 9th 1874."[69]

The trip did not end there. The Mexicans remained in Yokohama until the end of January, so they had close to two months to meander around the city and register their impressions. They spent most of their time attending diplomatic parties and personal meetings, participating in academic conferences, giving lectures in scientific institutions, and finishing their astronomic calculations.[70] On the night of February 2, the Commission left Yokohama on the French steamer *Tanais*, which

---

64. Bulnes, *Sobre el hemisferio norte*, 96–98, 129–30.
65. Díaz, *Viaje*, 118.
66. Moreno, *Odisea*, 80.
67. Díaz, *Viaje*, 19–21.
68. Moreno, *Odisea*, 82–92.
69. Cited in Moreno, *Odisea*, 93. The telegram was originally sent in English.
70. Moreno, *Odisea*, 98, 100.

took nine days to reach Hong Kong.⁷¹ In his usual sarcastic tone, Bulnes described the port as "a sterile rock where 700,000 human beings live and 300 million of opium get sold annually [sic]."⁷² He noticed a mixed population, of which "the Chinese, filthy as the conscience of a demon, occupy the lower part; the English, taken to those regions by the hurricane of speculation, have taken the elevated part of the mountain, where they have erected fortresses and have declared Hong Kong the first port of Asia."⁷³ Despite—or because of—the fact that both groups shared a certain "national pride and the intelligence for trade,"⁷⁴ they got along quite badly as "the Chinese ferocity and the subsequent reprisals of the foreigners [against them] maintain the social relations between both races in a state of permanent hostility."⁷⁵ The commissioners only remained in Hong Kong for two days as it was just a stopover on their long journey back to Mexico. Díaz Covarrubias had decided to publish the Commission's astronomic measurements in Europe first, rather than in Mexico, to gain a wider international audience and to avoid any possible confrontation with the oppositional press back home. Therefore, they took an even longer route to return, leaving Hong Kong aboard *Tigre*, a French steamer that did the long run between Shanghai and Marseille via the Suez Canal.⁷⁶ The scientists disembarked in Naples and slowly made their way to Paris by train. Once there, in the summer of 1875, the Mexicans became the first state-sponsored team to publish their results, followed by the French in 1877, the English in 1881, and the Russians in 1891.⁷⁷ Eventually the international scientific community would realize that none of the teams' measurements improved on the original, inaccurate seventeenth-century calculations—something that Díaz Covarrubias had sensed on site back in Japan.⁷⁸ Fortunately for them, nobody was aware of this when the public gave them a hero's welcome at the Mexico City railway station on November 19, 1875, nor later when they were received with honors by the ENP community.⁷⁹

---

71. Bulnes, *Sobre el hemisferio norte*, 180–81. According to Moreno, *Odisea*, 107, only Bulnes left aboard the *Tanais*. The other astronomers left on the British steamer *Volga*. They all met again in Hong Kong.
72. Bulnes, *Sobre el hemisferio norte*, 182.
73. Bulnes, *Sobre el hemisferio norte*, 182.
74. Bulnes, *Sobre el hemisferio norte*, 194.
75. Bulnes, *Sobre el hemisferio norte*, 190.
76. Bulnes, *Sobre el hemisferio norte*, 211.
77. All the other teams either published their results after the Russians or didn't do so at all. Moreno, *Odisea*, 119–20.
78. Moreno, *Odisea*, 97, 124. For a non-specialized article about the Astronomical Unit and how the distance between the Earth and the Sun is calculated, see Rebecca Sohn and Doris Elin Urrutia, "Astronomical Unit: How Far Away Is the Sun?," Space.com, last modified November 1, 2023, https://www.space.com/17081-how-far-is-earth-from-the-sun.html.
79. Francisco Jiménez was the only one who returned earlier on his own. Moreno, *Odisea*, 118–22.

Traveling Asymmetries

The memoirs left by Francisco Díaz Covarrubias and Francisco Bulnes allow us to determine not only the route and the vicissitudes of the trip described in the previous section, but also the ideological context and debates that would define the travel of large numbers of Chinese and Japanese to Mexico in subsequent years. There are at least three key contexts to consider when examining the social commentaries embedded in the texts. The first has to do with the importance given to positivist science and its embedded relationship to politics, explained earlier in this chapter. The second relates to the racialized prejudices that scientists held, not only in Mexico but wherever Europeans had colonized. In the context of the abolition of slavery and the slave trade and the expansion of imperial colonialism to Asia and Africa, theories that justified the domination of "inferior races" emerged.[80] Adapting ideas of evolutionary change from the primitive to the superior, prevalent since the seventeenth century, European scientists divided humanity into a hierarchy of racial types, each of which supposedly had innate qualities that were passed on to ensuing generations. At the top were whites, allegedly the most rational and civilized who, according to the utilitarian principle of common good, had the right—and even the obligation—to govern those deemed as inferior. While the exact hierarchy could vary, most situated blacks at the bottom and saw the so-called mix-bloods as degenerates. In general, Mexican—and Latin American—scientists situated themselves rather ambivalently within this paradigm. On the one hand, they accepted the premise of the existence of a hierarchy of "races." On the other hand, they rejected the notion of racial determinism and instead advocated for the possibility of improvement of racial traits. After all, if they accepted that mix-bloods were degenerate and that "Indigenous" or "Indians," the term they used to refer to people of First Nations, were innately inferior, there would be no promising future for Mexico, where the large majority of the population were deemed Indigenous or Mestizos.[81] If social conditions were favorable, Latin American scientists sustained, an inferior "race" could improve its qualities over time.[82]

The last major context is the relationship between travel/travel writing and empire- and nation-making. Both first-class travel and travel publishing increased in the nineteenth century with the advent of steam technologies. The economic prosperity of the middle and upper classes and their subsequent interest in travel

---

80. Peter Wade, *Race and Ethnicity in Latin America* (New York: Pluto Press, 2010), 10.
81. As mentioned in the Introduction, "Mestizo" is a complex and dynamic term, but in the memoirs analyzed in this chapter, the notion refers to a mixed ancestry from European and First Nations, where the former was more appreciated for its alleged superior qualities of economic and "racial" progress, as discussed later in this section. Chapter 3 will discuss the concept more thoroughly.
82. Wade, *Race and Ethnicity*, 10–31. See also Richard Graham et al., *The Idea of Race in Latin America, 1870–1940* (Austin: University of Texas Press, 1990); Nancy Stepan, *"The Hour of Eugenics": Race, Gender, and Nation in Latin America* (Ithaca, NY: Cornell University Press, 1991).

literature made the genre thrive.[83] The privileges that allowed wealthy travel writers to move freely influenced the way they understood and wrote about the places they visited, so they often reproduced racist, sexist, and other unequal discourses existing in their societies of origin. Their travel pieces in turn reinforced prejudices already present among their massive audiences.[84] In this sense, travelers who were sponsored by or at the service of governments, such as the Mexican scientists, were more likely to justify and reproduce the asymmetrical premises defended within their political circles in their texts. This was visible in travel writing conventions, where power structures were "replicated in textual patterns of signification and narrative authority. At one level, these [were] acquired and maintained through clear-cut binaries expressed in the narrative, such as superior culture/inferior culture, modernity/primitiveness, enlightenment/darkness, and scientific worldview/superstition. At another level, the patterns of signification reflect[ed] an orderliness based on: binarism; hierarchy; division of class, race, gender, and religion . . . This orderliness can be found in the type of narrative voice the travel writer chooses, as well as in the textual and figurative structure, and in the motifs, images, and metaphors that circulate in the text."[85] At the core of all these tropes was the ultimate binary of European modernity, widely defended by Latin American elites: civilization versus barbarism.[86] "Without civilization man is the most servile slave to nature,"[87] wrote Bulnes. And since he and his fellow travelers belonged to the "civilized" portion of humanity, superior to those who did not believe in paradigms of progress and development, they possessed the credentials to judge and make authoritative observations about "Others."[88] As such, travel narratives such as those from Díaz Covarrubias and Bulnes portrayed a regime of truth that was supposedly rational, scientific, and unbiased but that, more often than not, responded to literary romantic conventions full of drama, myths, and heroic actions, where contradictory statements were not uncommon, as will become evident in subsequent paragraphs.[89]

---

83. Shannon Marie Butler, *Travel Narratives in Dialogue: Contesting Representations of Nineteenth-Century Peru* (New York: Peter Lang, 2008), 2.
84. In her introduction to *Defining Travel: Diverse Visions* (Jackson: University Press of Mississippi, 2007), editor Susan Roberson provides an excellent synthesis of the state of the art in travel studies, including the ideas discussed in this paragraph. In addition, the book includes texts from twenty-one leading scholars.
85. Paul Smethurst, "Introduction," in *Travel Writing, Form, and Empire: The Poetics and Politics of Mobility*, ed. Julia Kuehn and Paul Smethurst (New York: Routledge, 2009), 6.
86. Daniar Chávez Jiménez, "Viajeros del siglo XIX: el linaje mexicano y las 11 mil leguas de Francisco Bulnes por el hemisferio norte," *Estudios* 108, no. 12 (Spring 2014): 64.
87. Bulnes, *Sobre el hemisferio norte*, 25.
88. Butler, *Travel Narratives*, 4; Ali Behdad, "The Politics of Adventure: Theories of Travel, Discourses of Power" in *Travel Writing, Form, and Empire: The Poetics and Politics of Mobility*, ed. Julia Kuehn and Paul Smethurst (New York: Routledge, 2009), 86.
89. Smethurst, "Introduction," 1. In the introduction to his book, Bulnes seemed to position himself differently by sustaining that "I do not intend to present a complete truth, nor to position myself at the height required by matters of scientific or moral speculation." Bulnes, *Sobre el hemisferio norte*, 7. But his text is full of tropes and writing strategies such as those described in this paragraph, as will be explained later, which actually suggest the opposite or, at the very least, highlight the contradictions often present in these travel texts, as observed by Mary Louise Pratt, "Scratches on the Face of the Country; or, What Mr. Barrow Saw in the

While the binaries and writing strategies described above proliferated among imperial and national emissaries, the latter's particular loci of enunciation made them both complicit and critical of imperialist discourses. In this sense, when Díaz Covarrubias and Bulnes talked about imperial centers, they did so with ambivalence. For the case of Spain, Díaz Covarrubias harshly criticized its colonial enterprise in the Americas: "Our conquerors, not cruel enough to destroy the subjugated race, nor generous and visionary enough to assimilate and elevate it to their stature by civilizing it, [rather] opted for the worst of the middle grounds, that is, to morally annihilate it." The politics of King Philip II (1556–1598) were "a terrible mixture of authoritarian zeal and religious fanaticism, inquisition, and dominion."[90] At the same time Díaz Covarrubias favored the arrival in Mexico of "laborious immigrants" from Spain and other parts of Europe "so that working for themselves they cooperate effectively to the prosperity of their new homeland."[91] Bulnes did not refer to the Spanish colonial enterprise openly, but when meeting Spanish servers in Havana, he was disdainful: "they crawl rather than walk, sweat like circus horses, and their intelligence is uncultivated."[92]

As for "the Anglo-American race," as Díaz Covarrubias labeled it, both writers showed a combination of fascination and disdain while traveling across the United States. For example, Bulnes sustained that "all American cities look alike, the houses are identical and the society is the same . . . It's not like with us . . . [in Mexico where] social diversity . . . causes an unconscious fight that impedes the prosperity that we must envy from our neighbours."[93] But later in his text, he criticized the people moving up and down according to routine schedules like "machines without a steering wheel . . . Without a doubt, the people here love civilization the most; but having invented the mysticism of mass production, [they] love it in blocks, without even remembering that refinement exists."[94] Díaz Covarrubias also showed contempt for "disgusting Anglo-American habits,"[95] such as those described by Bulnes, but, like him, he was full of admiration for their economic growth and technological advancements, as shown in his praise of their "superior" trains and "magnificent" cities and especially when he referred to the English-speaking doctors traveling on *Vasco de Gama* to Japan and China who, for him, were going "to distribute amongst

---

Land of the Bushmen," in *Defining Travel: Diverse Visions*, ed. Susan Roberson (Jackson: University Press of Mississippi, 2007), 135. For a concise and interesting analysis on Bulnes's style, possible contradictions, and additional topics to those addressed in this book, see José Ricardo Chaves, "Estudio preliminar: Bulnes viajero," in *Francisco Bulnes. Sobre el hemisferio norte once mil leguas. Impresiones de viaje a Cuba, los Estados Unidos, el Japón, China, Conchinchina, Egipto y Europa. Edición facsimilar* (Mexico City: UNAM, 2012), 7–26.

90. Díaz, *Viaje*, 26.
91. Díaz, *Viaje*, 41.
92. Bulnes, *Sobre el hemisferio norte*, 23.
93. Bulnes, *Sobre el hemisferio norte*, 87–88.
94. Bulnes, *Sobre el hemisferio norte*, 46. For an interesting compilation on Mexican travelers to New York between 1830 and 1895, including a short analysis of Bulnes's trip to this city, see Vicente Quirarte et al., *Republicanos en otro imperio. Viajeros mexicanos a Nueva York (1830–1895)* (Mexico City: UNAM, 2009).
95. Díaz, *Viaje*, 77.

these [Asian] peoples the benefits of Western scientific progress, almost unknown to them."[96]

The texts described nature in various sections. The presence of wind measurements was frequent whenever they were navigating, especially as the commissioners traveled with a barometer. Nature was also described as a threat that hampered the Commission's possibilities of success as shown in earlier quotes, or in relation to its potential for being tamed and used for creating economic wealth.[97] For instance, when peeking out the windows of the transcontinental train traveling to San Francisco, Díaz Covarrubias seemed favorably surprised by the "abundance of navigable rivers . . . [which to him explained why] our sister Republic had prospered so quickly . . . If we could only exchange all the gold and silver mines that make Mexicans so proud for all these extensive rivers and lakes that incite the population . . . to movement, trade, and to be in contact with the most remote peoples."[98] Nevertheless, his perspective turned sour once the train reached the Rocky Mountains, where Díaz Covarrubias noticed how the abundant snow made it necessary to build numerous sheds to protect the tracks. He noted that these costly structures would often burn with the heat created by the locomotives, making the whole train operation expensive and inefficient. "All these inconveniences, along with the decisive factor of the enormous distance,"[99] made Mexico's Tehuantepec isthmus a far superior option for transcontinental trips. The construction of a railroad in Tehuantepec, the narrowest stretch in Mexican territory between the Atlantic and Pacific oceans, was a constant topic in political discourses at the time and would eventually prove important for transoceanic travel, as will be explained in Chapter 5.

Both memoirs contain numerous prejudiced comments reinforcing racialized hierarchies among various ethnic groups. These began in Havana, their first stop, where both Díaz Covarrubias and Bulnes noted a "racial" heterogeneity and antagonism: "population [is] composed . . . of a third of whites . . . and two thirds of Africans and their varied racial mixtures; antipathy is deeply rooted one against the other."[100] Bulnes was disgusted by how "blacks flaunt their filthy misery and wander around the city with the heavy step of prisoners."[101] Díaz Covarrubias, on his part, wrote that during one of the transcontinental train's meal stops he wandered around and fell attracted to a "savage of athletic height, [with] very pronounced and not disagreeable factions . . . a meditative attitude, [and an] arrogant and sad look over

---

96. Díaz, *Viaje*, 95.
97. In relation to this, Pratt notes that "nineteenth century exploration writing rejoins two planetary processes that had been ideologically sundered: the expansion of the knowledge edifice of natural history and the expansion of the capitalist world system. . . . In scanning prospects in the spatial sense—as landscape panoramas—this eye knows itself to be looking at prospects in the temporal sense—as possibilities for the future, resources to be developed, landscapes to be peopled or repeopled by Europeans." Pratt, "Scratches," 138.
98. Díaz, *Viaje*, 76.
99. Díaz, *Viaje*, 80.
100. Díaz, *Viaje*, 53.
101. Bulnes, *Sobre el hemisferio norte*, 22.

the crowd [that] revealed the most absolute indifference and lack of curiosity."[102] He tried to engage him in conversation by giving him some coins, to which the man replied with a hostile look that the Mexican interpreted as a "deep aversion to civilized men." This incident served him to reflect on First Nations, labeled by him as the "American race": "Science will one day explain the physiological causes that determine the degree of perfection of each race and that, at present, we can only appreciate by its effects. . . . None of the aboriginal peoples of the Americas have learnt not even to emancipate and to put themselves on a par with other nations, but also to at least learn something of the civilization of their oppressors."[103]

Bulnes and Díaz Covarrubias wrote more benevolently about the Japanese, but still with the usual authoritative distance and invoking the paradigms of progress that characterized travel writing as explained above. Bulnes said of the Japanese: "they sleep little and work continually. They have a sweet character, love foreigners and are always ready to serve them. They never ask to be remunerated for their work, but rather wait until they are given something and whether they receive little or much, show their gratitude."[104] But he also made a sarcastic remark on their exaggerated work ethic: "another Flood might come and all the snow of the poles might fall and the Japanese would not stay still at home."[105] Díaz Covarrubias, on his part, described them as "poor and laborious . . . sober, provided by [their] education with a spirit of order and respect for the law, accustomed to look for subsistence through work."[106] In his view, these traits would make them an adequate asset if they immigrated to Mexico because they "would provide our landlords with a great number of cheap, active, and intelligent laborers; at the same time a Japanese colony would offer a healthy example to our people of everything that can be achieved with perseverance, hard work, and economy, even in the most unfavorable conditions."[107]

In contrast, the Mexicans' remarks on Chinese were, for the most part, hostile and discriminatory. Bulnes described them as "filthy"[108] and found their customs, in the limited days he spent in Hong Kong, notably inferior to those of the Japanese. He added: "They believe themselves to be superior to all nations and consider their civilization, not only the eldest, but also the most brilliant. Their tendency to pillage makes them terrible. . . . Their classic and well-earned reputation for cruelty is the consequence of their cowardice and rapaciousness [is] the glory of their misery."[109] Díaz Covarrubias shared the same contempt for Chinese, whom he defined as "sly, distrustful, and full of instinctive aversion towards everything but their motherland. They perform the work just for the love of profit . . . [and] execute the labor assigned

---

102. Díaz, *Viaje*, 83.
103. Díaz, *Viaje*, 84.
104. Bulnes, *Sobre el hemisferio norte*, 121.
105. Bulnes, *Sobre el hemisferio norte*, 121.
106. Díaz, *Viaje*, 129.
107. Díaz, *Viaje*, 129.
108. Bulnes, *Sobre el hemisferio norte*, 22, 182.
109. Bulnes, *Sobre el hemisferio norte*, 194.

to [them] as a machine, without becoming fond of it. . . . In the meantime, they cherish the dollars that they hide perhaps under the folds of their tunic, patiently waiting for the moment to enjoy them back in their homeland."[110] This observation of their active nature contrasted with a later description of Chinese as essentially stationary, in opposition to the progressive nature of the Japanese people.[111] All this was despite the fact that their only meaningful and extended interaction with a Chinese person, the carpenter who built their observatories in Yokohama, was described, as we saw earlier, in highly positive terms.[112]

Bulnes and Díaz Covarrubias's prejudices against Chinese were informed by a larger discussion that had started in October 1871 in the Mexican press, following the arrival in the port of Veracruz of a large group who had been expelled from Havana. Cuba, still a Spanish colony, was one of the American territories where Chinese contract laborers had had a widespread presence since the mid-century. In the context of a booming sugar industry, the pillar of the island's economy, Spanish planters brought in Chinese farmers to supplement African slave labor, whose trade was then on the verge of elimination.[113] Between 1847 and 1874, the years that the so-called coolie trade officially existed,[114] 347 vessels brought some 125,000 male contract workers to Cuba. While, in theory, they traveled according to their free will with a signed contract that stipulated a payment in exchange for their work, in reality many were forced or misled with false promises and lived in slave-like conditions.[115] At the same time, there were many who, after fulfilling their contracts and pooling their savings with those of friends and family, were able to become prosperous, independent businessmen. The contingent that left Cuba for Mexico in 1871

---

110. Díaz, *Viaje*, 82.
111. Díaz, *Viaje*, 126–27.
112. In terms of how these visions intersect with discussions in the neighboring American countries, notably the United States, see Robert Chao Romero, *The Chinese in Mexico, 1882–1940* (Tucson: University of Arizona Press, 2010); Grace Peña Delgado, *Making the Chinese Mexican: Global Migration, Localism, and Exclusion in the U.S.-Mexico Borderlands* (Stanford, CA: Stanford University Press, 2012); Young, *Alien Nation*; Erika Lee, *The Making of Asian America: A History* (New York: Simon & Schuster, 2015); Jason Oliver, *Chino: Anti-Chinese Racism in Mexico, 1880–1940* (Champaign: University of Illinois Press, 2017); Karin Alejandra Rosemblatt, *The Science and Politics of Race in Mexico and the United States, 1910–1950* (Chapel Hill: University of North Carolina Press, 2018).
113. Cuba and Brazil were the last two American territories to abolish slavery in the 1880s. Walton Look Lai, "The Caribbean," in *The Encyclopedia of the Chinese Overseas*, ed. Lynn Pann (Cambridge, MA: Harvard University Press, 1998), 248.
114. The term coolie most probably came from an Indian word meaning servant and was then used to refer to Chinese laborers who traveled abroad with a fixed three- to eight-year contract. Gómez, *Movimiento antichino*, 29.
115. They normally signed (or were forced to sign) eight-year contracts that stipulated a one-peso weekly salary. The workers were supposed to be provided with sufficient food, clothing, medical services, housing, and one free day per week, plus three days off during New Year. They normally received between eight and fourteen pesos at the time of departure to pay for their passage. This amount was owed to their bosses and paid by deductions from their salaries at the rate of one peso per month. Evelyn Hu-DeHart, "On Coolies and Shopkeepers: The Chinese as Huagong (Laborers) and Huashang (Merchants) in Latin America/Caribbean," in *Displacements and Diasporas: Asians in the Americas*, ed. Wanni W. Anderson and Robert G. Lee (New Brunswick, NJ: Rutgers University Press, 2005), 85.

was part of this latter trend. They had to depart after the Spanish authorities enacted a Royal Order that suspended Chinese immigration to the island and expelled those whose contracts had expired. This was done in the context of a Cuban insurrection seeking independence from Spain that was gaining increasing support from the local Chinese population.[116]

After their arrival in Mexico on October 12, the federal government's *Diario Oficial* summoned the media to reflect upon the repercussions of attracting the twenty to thirty thousand Chinese who might be expelled from Cuba. Without taking into consideration that most of the new arrivals were retail businessmen, the *Diario* suggested that they could boost the Mexican economy by providing cheap, hardworking, and submissive labor for those areas with harsh geographic conditions as well as for the construction of the railroads. This opinion was shared, for instance, by the newspaper *El Federalista*. In an article titled "The Chinese Immigration," its director, Alfredo Bablot, distinguished three types of immigrants: the capitalist, the *hombre de genio* (creative genius), and the laborer. While the first two had enough material and intellectual talent to start successful economic ventures, Bablot thought they would merely increase their own individual profits without much regard for the national interest. The latter category—of which Chinese formed part, according to him—was composed of simpleminded, hardworking, and submissive individuals who, therefore, could be manipulated to "serve as an instrument for the material exploitation of any of the industries [that create] public wealth."[117] The opposition newspaper, *El Siglo XIX*, had a different perspective. An article written by federal deputy Jesús Castañeda summarized the main objections given by all those who opposed Chinese immigration in the press. He claimed that Chinese were "the least civilized people: used to misery and dominated by greed, they deny their body of the advantages of a comfortable and hygienic life . . . their favorite vices [are] gaming and intoxication . . . their small and unfurnished rooms accommodate a considerable number of guests of both sexes, turning those smoky and greasy houses into disgusting pigsties, where all that can be repugnant in their unbridled customs nest."[118] What both seemingly opposing perspectives shared was a view of Chinese as a foreign working class. No one in the press ever discussed their integration into society in different terms nor highlighted the ancestral roots that had linked Chinese people with Mexico since the circulation of the galleons.

Conclusions: Mexico and Asia in the 1870s

Díaz Covarrubias's and Bulnes's memoirs of their trip to Asia aboard *Vasco de Gama* have enabled us to determine the travel routes and conditions as well as to discuss

---

116. Gómez, *Movimiento antichino*, 44–45.
117. Gómez, *Movimiento antichino*, 48; *El Federalista*, October 24 and 31, 1871.
118. Gómez, *Movimiento antichino*, 46; *El siglo XIX*, October 24, 1871.

some of the historical processes and discourses that shaped the transpacific flow of peoples and ideas to and from Mexico in the 1870s. Their opinions were representative of a Mexican political class that sought to consolidate their positivist and developmentalist projects by promoting the immigration of cheap laborers. Their writings also formed part of a series of asymmetrical travel writing conventions and travel conditions that favored first-class passengers. When put into context, the memoirs highlighted the ways in which the interlinked notions of imperialism, race, class, nation-making, travel, and science framed the uneven experiences of passengers between Mexico and Asia, including those discussed later in this work.

In terms of the transpacific route and services linking Mexico and Asia in the 1870s, the memoirs confirmed many of the observations discussed in Chapter 1, such as the predominance of PMSS's steamships and of San Francisco as a stopover as well as the centrality of Chinese steerage passengers, whose fares were the main business that sustained transpacific runs. In comparison, first-class passengers were few and came mostly from the United States and Europe, with the odd wealthy travelers from other countries, such as the five Mexican scientists. This hierarchy and these numbers were similar for the ships' crew, where the few high officers were from the United States and the numerous sailors and other lower posts were Chinese. The contextualization of the scientists' trip also showed that during this decade, the Mexican government began a policy of subventions of foreign steamship companies to exert some control over the traffic and to remain connected with US and Central American ports. In the absence of railroads, the Mexican Pacific coast remained relatively isolated by land from the rest of the country.

In terms of the social commentaries found in the memoirs, they reproduced the kinds of tropes and contradictions present in travel and scientific writings of the time, where the authors positioned themselves as an expert voice that reinforced previous assumptions on racialized hierarchies. Díaz Covarrubias and Bulnes did so by resorting to binaries such as civilized/savage, progressive/stationary, and rational and scientific/fanatic whenever they described the people they encountered. Their opinions as travelers were even more influential as they were also scientists working for one of Mexico's most revered educational institutions, the ENP. In this context, scientific and traveling assertions highly valued by a wide audience merged to create a privileged discourse that supported and helped spread racist attitudes.

As both memoirs came from a state-sponsored trip, their authors regularly used the social commentaries previously analyzed to propose developmentalist solutions for Mexico in tune with the racialized asymmetries prevalent in political circles and travel writings of the time. Their key initial premise was that the majority of Mexico's population, pertaining to what Díaz Covarrubias labeled the "American race,"[119] was for him "one of the most useless for the progress of humanity."[120] As a

---

119. In his context, this refers to First Nations.
120. Díaz, *Viaje*, 84.

long-term "solution" for this "problem," he proposed "the education of the masses ... which is the only element capable of establishing, up to a certain point, equality amongst men."[121] But in the short term, he thought it urgent to promote abundant immigration.[122] For this purpose, he favored some "races" more than others, notably Europeans who could hopefully "mix" with the locals so that future generations could inherit the qualities of the most "rational and civilized" part of the equation.[123] This, for him, could also serve to create a supposedly more homogeneous "Mexican race" or Mestizo, as he viewed the alleged racial heterogeneity present in the country as an obstacle to progress and a potential source of conflict.

Of the different topics addressed in the memoirs, the most relevant for subsequent chapters is the discussion of Chinese and Japanese, often framed in developmentalist terms where their only position within the nation was as laborers. Influenced by earlier debates in the Mexican press following the arrival of a sizable group of Chinese merchants fleeing Cuba, both scientists wrote mostly negative comments about Chinese, which are reminiscent of the dismissive stereotypes listed by Spanish colonial authorities described in Chapter 1. This was despite forming a positive impression of the only Chinese person with whom they actually engaged. In contrast, they described Japanese as courteous, hardworking, obedient, and progressive because of their embrace of Western scientific progress, which was something they desired for Mexico. The distinction made between Chinese and Japanese and the association of the latter with an alleged impetus for progress would be a recurrent trope discussed by Mexican elites, but opinions on the most ideal type of immigration also diverged, as will be examined in the subsequent chapters.

While the international scientific community failed at calculating the distance between the Sun and the Earth by measuring the transit of Venus in 1874, the Mexican Astronomic Commission's journey around the world succeeded in its political objectives, enhancing the prestige of the government of President Lerdo de Tejada. Time would eventually show, however, that Lerdo needed much more than the help of Venus to remain in power. Two years after the Commission's journey, Lerdo de Tejada was living in exile in New York. In December 1876, a military coup led by General Porfirio Díaz Mori ousted him from the presidency. The repercussions for Mexican transpacific relations of this change of power are the focus of the next chapter.

---

121. Díaz, *Viaje*, 53.
122. Díaz, *Viaje*, 41.
123. Díaz, *Viaje*, 41.

3
Mount Lebanon, 1884
*Navigating in Britain's Diplomatic Waters**

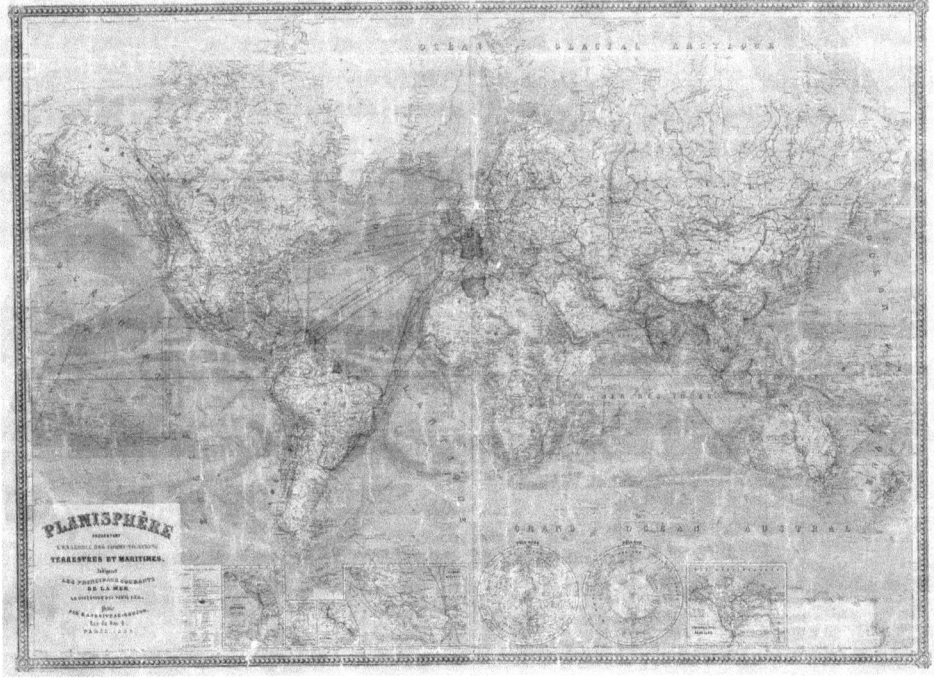

Map 3
Vías de comunicación marítimas y terrestres [Maritime and land routes]. Map. Mexico City, E. Andriveau Coujon, 1882. From Mapoteca "Manuel Orozco y Berra," Planisferios 1. Accessed December 15, 2023. https://mapoteca.siap.gob.mx/cgf-planf-m35-v1-0064/.

---

* A previous version of this chapter was published in the *International Journal of Maritime History*. Reproduced here with the permission of the journal's Rights Administrator Craig Myles.

On the morning of October 6, 1884, the Mexican Foreign Secretary, José Fernández, received an urgent message in his Mexico City office. It stated, "the [British] Governor of Hong Kong [was] not willing to allow [Chinese laborers] to emigrate to this Republic."[1] The signers were Salvador Malo, Guillermo Vogel, and Luis Larraza, owners of the Compañía Mexicana de Navegación del Pacífico (CMNP), with whom the Mexican government had recently signed a contract to "establish maritime routes between Mexico and Asia" in order to bring in "Asian laborers and European immigrants."[2] The company's inaugural transpacific steamer, *Mount Lebanon*, had arrived in Hong Kong a few days earlier and was now ready to depart for Mexico. Rather than do so without the desired passengers, the owners requested the secretary's immediate intervention.

The note could not have arrived at a more delicate time. At that moment, the Mexican Senate and British Parliament were examining the Preliminary Agreement for the renewal of bilateral relations, broken in 1862 at the height of strain between Mexico and Europe over the French invasion discussed in Chapter 2.[3] A faux pas could compromise negotiations at a time when the Mexican government was particularly eager to re-establish diplomatic ties. The country had recently held presidential elections, won by General Porfirio Díaz Mori, and those involved in the transition wanted to gain foreign credibility and financial support for their modernizing projects.[4]

In their respective London and Washington offices, Mexican representatives Ignacio Mariscal and Matías Romero soon found themselves involved in the issue. Mexico had no diplomatic relations with China, so the only way to have direct contact with its officials was through its representatives in London and Washington, Marquis Tseng Chi-Tse and Cheng Tsao Ju respectively. Mariscal also spoke to the Earl of Derby, the British Secretary of State for the Colonies. Despite courteous replies, no immediate action was taken.[5]

Mount Lebanon, on its part, had had fine weather throughout its twenty-day crossing from Australian waters to Hong Kong.[6] The 1,555-ton iron steamer, built in A. Stephen & Sons' Glasgow yards three years prior,[7] was now CMNP's first and

---

1. Archivo Histórico Genaro Estrada, Secretaría de Relaciones Exteriores (AHGE-SRE), 44-6-35, 1. Translations from this archive's documents into English were done by the author.
2. AHGE-SRE, 44-6-35, 102–3.
3. The government of Benito Juárez broke off relations with all European countries that recognized the French-backed Maximilian of Habsburg's government, including England. Official bilateral relations had not been renewed since then. On this subject, see Konrad Ratz, *Tras las huellas de un desconocido: nuevos datos y aspectos de Maximiliano de Habsburgo* (Mexico City: Conaculta/Siglo XXI, 2008); Josefina Zoraida Vázquez, ed., *Interpretaciones del periodo de Reforma y Segundo Imperio* (Mexico City: Patria, 2007).
4. See *Further Correspondence Respecting the Renewal of Diplomatic Relations with Mexico* (London: Foreign Office, 1885), 1–50.
5. Vera Valdés Lakowsky, "México y China: del Galeón de Manila al primer tratado de 1899," *Estudios de historia moderna y contemporánea de México* 9 (1983): 9–19.
6. *The China Mail*, October 3, 1884, 4–5.
7. *Lloyd's Register of British and Foreign Shipping* (London: Wyman and Sons, 1881), MOU.

only vessel, chartered from another British company, Smith & Service. For weeks, the company had advertised in *The China Mail*—Hong Kong's most prominent English newspaper—that it would "run monthly steamers . . . from Hong Kong to Yokohama and Honolulu; thence to Mazatlan (direct), and from Mazatlan to San Blas, and Manzanillo."[8] However, while attempting to embark Chinese laborers destined for Mexico, the British governor of Hong Kong refused to let them depart to a country that had relations with neither England nor China.[9]

Months of negotiations among stakeholders ensued. Their analysis reveals the numerous variables involved in the reconfiguration of Pacific maritime relations after the introduction of steam technologies and the participation of Mexican interests in this process. Contextualizing the case exposes the importance that Asian immigration and foreign relations with China played in Porfirio Díaz's earlier mandates. It also shows how the ideas debated in the 1870s, discussed in Chapter 2, evolved and affected Cantonese travelers to Mexico in the following decade, when neighboring American governments began to impose restrictions on Chinese mobility.

Porfirian Modernization

1876 marks the beginning of the Porfiriato,[10] the longest presidency ever held by a single person in the history of Mexico. General Porfirio Díaz Mori would rule the country until 1911 with the exception of a four-year period between 1880 and 1884, when his long-time friend and collaborator General Manuel González took the reins.[11] Díaz began his political and military career as a close associate of President Benito Juárez. Battlefield success against the French intervention won him the support of a broad swath of war veterans; however, it would take yet more confrontation to get voters on his side. The general contended in the 1870 presidential elections and lost. Claiming fraud, he initiated an armed revolt that was ultimately suppressed by Juarista forces. He then ran in the 1872 elections after Juárez's death but came second to Sebastián Lerdo de Tejada. During this mandate, Lerdo was accused of manipulating various state elections and placing close associates in power. Opposition mounted, even among liberals, when he announced

---

8. *The China Mail*, October 3, 1884, 5.
9. AHGE-SRE, 44-6-35, 1.
10. The Porfiriato has been the subject of multiple analyses. For the main historiographical debates on the subject, see Paul Garner, *Porfirio Díaz* (London: Pearson, 2001), 1–17, and Mauricio Tenorio Trillo and Aurora Gómez Galvarriato, *El Porfiriato* (Mexico City: CIDE/FCE, 2006), 23–95.
11. Most researchers have treated the presidency of General Manuel González as part of the Porfiriato, due to his close connections to Díaz, both personal and political—Díaz formed part of González's Cabinet as Secretary of Development and President of the Supreme Court, which, according to the Constitution, gave him the presidential succession. Yet Paul Garner and other researchers have been cautious to suggest that González was merely a puppet of Díaz. Garner, *Porfirio Díaz*, 90–94. For the purpose of this chapter, it makes sense to treat González's presidency as part of the Porfiriato because there were no major disruptions when it came to the policies discussed here.

his candidacy in 1876. Alluding to the violation of the constitutional principle of popular sovereignty and the concentration of illegal powers (charges of which Díaz would eventually be accused), the general rallied a new armed revolt against Lerdo's re-election attempt, which was ultimately successful.[12]

Once in power, Díaz faced political isolation from the international community. The United States refused to recognize his government as it came via a *coup d'état*.[13] Even after legitimizing his position by calling for—and winning—elections at the beginning of 1877, President Rutherford B. Hayes remained obstinate, attempting to obtain more concessions.[14] Across the Atlantic, the situation was similarly bleak. Juárez had broken off relations with European countries that had recognized the monarchical government of Maximilian of Habsburg. By the time Díaz became president, only Spain had resumed ties.[15] In this context, Porfirian diplomacy sought to normalize relations with the industrialized world. In 1878 it finally obtained White House recognition.[16] A year later it resumed ties with Belgium,[17] followed by France in 1880.[18] In 1882 Germany signed a bilateral treaty.[19] By 1884, when the *Mount Lebanon* controversy erupted, only one major economic power was missing: Great Britain.[20] Since its independence from Spain, Mexico had never had any diplomatic relations with Asian countries.

Domestically, Díaz's modernizing program consisted in general terms of "four major components: (1) the establishment of political order and stability; (2) the recruitment of foreign capital and investment; (3) the creation of an extensive transportation network; and (4) the promotion of European immigration."[21] Economically, his governments managed to achieve the first period of sustained growth in the country's history,[22] albeit at the expense of political liberties and social

---

12. On Díaz's early political career, his confrontations with Juárez and Lerdo, and his succession to the presidency, see Garner, *Porfirio Díaz*, 48–67.
13. María de Jesús Duarte Espinosa, *Frontera y diplomacia: las relaciones México-Estados Unidos durante el Porfiriato* (Mexico City: SRE, 2001), 19.
14. Duarte, *Frontera*, 20; Silvio Zavala, *Apuntes de historia nacional, 1808–1974* (Mexico City: FCE, 1990), 120.
15. For a history of the relations between Spain and Mexico during this period, see Adriana Gutiérrez Hernández, "Juárez, las relaciones diplomáticas con España y los españoles en México," *Estudios de historia moderna y contemporánea de México* 34 (July–December 2007): 29–63.
16. Duarte, *Frontera*, 20; Garner, *Porfirio Díaz*, 177.
17. For a history of the relations between Belgium and Mexico, see Florence Loriaux et al., *Les Belges et le Mexique: dix contributions à l'histoire des relations Belgique-Mexique* (Louvain: Presses Universitaires de Louvain, 1993).
18. Paul Garner, *Porfirio Díaz: entre el mito y la historia* (Mexico City: Crítica, 2015), 337.
19. For a summary on Mexico-Germany relations, see María del Pilar Escobar Bautista, "México–Alemania: datos de una valiosa relación histórica," *Revista mexicana de política exterior* 99 (September–December 2013): 175–83.
20. For an analysis of the different phases of Porfirian diplomacy, see Garner, *Porfirio Díaz, entre el mito*, 201–34.
21. Robert Chao Romero, *The Chinese in Mexico, 1882–1940* (Tucson: University of Arizona Press, 2010), 25. See also Robert M. Buffington and William E. French, "The Culture of Modernity," in *The Oxford History of Mexico*, ed. Michael C. Meyer and William H. Beezley (New York: Oxford University Press, 2000), 398–400.
22. During its first fifty years of independence, Mexico had experienced continuous recession due to a combination of civil war, foreign invasions, lack of infrastructure and markets, failure of domestic policies, as well as a decline in mining. This situation took a remarkable turn in the last decades of the century, when its macroeconomic performance improved markedly, to the point that, in 1896, the country obtained its first ever surplus.

equality.²³ On the one hand, Mexican macroeconomics developed during a period of worldwide economic expansion related to industrialization,²⁴ when Latin American economies were incorporated into global markets as suppliers of raw materials and/or agricultural commodities and as importers of capital, technology, and, in some cases, labor.²⁵ On the other hand, the regime's longevity, stability, centralized power, and strong arm allowed for what its predecessors had been unable to guarantee: the continuity of federal policies. Important growth factors included the enlargement and diversification of industrial production, mainly in iron, steel, textiles, paper, glass, tobacco, and beer; the expansion of commercial agriculture, especially that of sisal; and the expansion of domestic and overseas trade, especially in mining and later in oil.²⁶ While the government promoted a variety of agricultural and industrial products, metals—and silver in particular—remained the most valuable export commodity throughout the Porfiriato.²⁷ As stated in Chapter 1, silver had been central to the colonial economy and its relations with China, but since the Mexican War of Independence, production had plummeted, mainly due to the loss of capital investment. Porfirian elites implemented a series of measures to stimulate mineral production.²⁸ As a consequence, numerous mining centers flourished, particularly in the north,²⁹ and silver exports increased from 607,000 kilograms in 1877–1878 to 2.3 million kilograms by the end of the Porfiriato in 1910–1911. By then, silver accounted for a third of Mexico's exports.³⁰

See John Coatsworth, "Obstacles to Economic Growth in Nineteenth-Century Mexico," *American Historical Review* 83 (1978): 80–100; Garner, *Porfirio Díaz*, 171.

23. Chapter 6 will touch on this aspect in relation to transpacific relations. For those interested in this broad topic beyond the scope of this book, see Garner, *Porfirio Díaz: entre el mito*, 279–314; Javier Garciadiego, "Aproximación sociológica a la historia de la Revolución Mexicana," in *Textos de la Revolución Méxicana* (Caracas: Fundación Biblioteca Ayacucho, 2010), 9–38; Timo Schaefer, *Liberalism as Utopia: The Rise and Fall of Legal Rule in Post-Colonial Mexico, 1820–1900* (New York: Cambridge University Press, 2017), 161–215.
24. Between 1853 and 1872, world trade grew at an unprecedented rate of 4.3 percent per annum and then at a still outstanding annual rate of 3 percent between 1872 and 1913. William Glade, "Latin America and the International Economy, 1870–1914," in *The Cambridge History of Latin America*, Vol. 4, ed. Leslie Bethell (Cambridge: Cambridge University Press, 1989), 8.
25. See Glade, "Latin America," 1–56.
26. Garner, *Porfirio Díaz*, 173–74.
27. Garner, *Porfirio Díaz*, 180–81.
28. Among these were the exemption and/or decrease in taxes for mining companies and their production supplies, the proliferation of mining concessions to foreigners, the decline in export duties, and the reduction of transportation costs inside the country. See Marvin D. Bernstein, *The Mexican Mining Industry, 1890–1950* (New York: State University of New York Press, 1964). Additionally, various mining chambers emerged, which, together with new government-sponsored associations, unified to lobby in favor of the industry holders. See Juan Manuel Romero Gil, *La minería en el noroeste de México: utopía y realidad 1850–1910* (Mexico City: Universidad de Sonora/Plaza y Valdés, 2001), 16–21, 107–13; Cuauhtémoc Velasco Ávila et al., *Estado y minería en México 1767–1910* (Mexico City: FCE/Secretaría de energía, minas e industria paraestatal, 1988), 321–55.
29. See the map "Mining centers and railroads, 1880–1910," in Velasco, *Estado y minería*, 256–57. For a study of one such region during the Porfiriato, see William E. French, *A Peaceful and Working People: Manners, Morals, and Class Formation in Northern Mexico* (Albuquerque: University of New Mexico Press, 1996).
30. Glade, "Latin America," 16.

A group of self-proclaimed "conservative-liberal"[31] intellectuals helped guide the philosophical pillars of productivity and economic progress to the forefront of Porfirian policies. Later known as the "Científicos,"[32] they comprised a generation of well-positioned pragmatists and included Francisco Bulnes—the Mexican Astronomic Commission's youngest member, discussed in Chapter 2. Influenced by the positivist notions disseminated by Gabino Barreda and his disciples from the Escuela Nacional Preparatoria (ENP),[33] the conservative-liberals believed that just as with nature, society should also be studied and ruled by scientific principles. Furthering the discussions that had taken place since the ENP's foundation, they defended the idea of progressive stages of thought that started with the theological or imaginary, advanced to the metaphysical or abstract, and culminated with the scientific or positive. They saw the liberal Constitution of 1857 as having advanced Mexican politics from the theological—represented by monarchical and imperial forms of government—to the metaphysical, as it now contained a series of abstract principles, such as the long list of universal individual rights. Yet they regarded these rights as exaggerated, arbitrary, and based more on faith than experience and science. It was thus time, they claimed, to move political thought a stage further by creating laws that emanated from the observable needs of Mexican society. After decades of anarchy, war, and generalized poverty, what the country needed most, according to them, was not abstract principles but material wealth. "We are tired of principles that do not represent the economic conscience of society," synthesized Bulnes. Instead, he advocated for a government "that develops the public wealth as quickly as possible."[34] Such a government should be led by those sectors of society that were the most productive, that is, the "industrialists,"[35] those who had capital and made it multiply through commercial, industrial, and agricultural ventures; those who, instead of armed revolutions, advocated for peace so that everyone could work and produce more. Contrary to the principles of the Juárez period, Bulnes's generation did not appeal to the principle of private property to create a nation of small holders. Instead, it supported a strong elite that could generate

---

31. The term "conservative-liberals" was coined in May 1878 in the newspaper *La Libertad*, edited by members of this new generation of intellectuals. Charles Hale, *The Transformation of Liberalism in Late Nineteenth-Century Mexico* (Princeton, NJ: Princeton University Press, 1989), 35.
32. The term "Científicos" (Scientists) entered the Mexican political jargon in 1893. It was given to the group of deputies and high-ranked officials linked to the notion of "scientific politics" discussed in this paragraph. In 1893, this group of politicians advocated for constitutional reform and supported president Porfirio Díaz's re-election. See César Arturo Velázquez Becerril, "Intelectuales y poder en el porfiriato: una aproximación al grupo de los científicos, 1892–1911," *Revista Fuentes humanísticas* 22, no. 41 (December 2010): 7–23.
33. See Chapter 2.
34. Cited in Richard Weiner, *Race, Nation, and Market: Economic Culture in Porfirian Mexico* (Tucson: University of Arizona Press, 2004), 28–29.
35. The term was coined by French positivist Henri de Saint-Simon and was sometimes used within Mexican positivist circles.

peace, order, and overall material progress—a social structure of which, of course, the conservative-liberals were already at the top.[36]

Within this train of thought, a central "problem" still to solve was "race"—the large majority of Mexico's population, defined in positivist racialized terms, was composed of "Indians" who were by no means the industrialist prototype.[37] As we saw, in his memoir, astronomer Francisco Díaz Covarrubias condemned both the Spanish colonizers and the Indigenous populations. Instead, some conservative-liberals—interested in reconciling the country's turbulent past to create a unified national identity and generate the conditions for peace and material progress—conceived a historical narrative to vindicate both the European and Indian pasts. They did so by fusing them into what was supposed to be the epitome of the Mexican nationality, the "Mestizo race." Drawing on social Darwinism and the positivist notion of successive stages leading towards progress, intellectuals championed by Vicente Riva Palacio, Secretary of Development in Díaz's first term, envisioned each historical stage as a necessary step towards its inevitable culmination: a liberal republic with a unique racial Mestizo composition, a product of its specific historical circumstances with legitimate claims to an equal status in and among the concert of nations. This notion was first elaborated in an extensive way in the largest historical synthesis ever undertaken to that point, *Mexico a través de los siglos* (Mexico throughout the centuries), the initial installment of which had been published a year prior to the *Mount Lebanon* controversy.[38] The reasoning defended in this work was not without its contradictions as the Spanish part of the Mestizo equation was valorized more and, while the intellectuals idealized the pre-Hispanic "Indian," they dismissed present-day Indigenous people due to their alleged incapacity for material progress, measured in positivist terms. Yet they tended to agree on the possibility of their "redemption."[39]

---

36. Weiner, *Race*, 22–31.
37. As discussed in Chapter 2, Western global elites supported the widespread idea of a human population composed of different "races" ordered hierarchically, where those considered superior shared the associated values of civilization, scientific rationality, and material progress. European positivism believed that biology transmitted positive or negative traits; subsequently, the more races mixed, the more degenerated their offspring became. The Mexican positivists—perhaps due to their existential conundrum—favored the possibility of racial improvement under the right conditions, such as positivist education and miscegenation with industrialist "races." This will be more thoroughly explained in the next paragraph.
38. The project, coordinated by Riva Palacio himself, consisted of five tomes, each devoted to a different historical period, starting with the "History of Antiquity," followed by the "History of the Viceroyalty," the "War of Independence," "Independent Mexico," and "The [Liberal] Reform," which ended in 1867, the year of the defeat of the French monarchist intervention. The five volumes are available in "México a través de los siglos," Biblioteca Virtual Miguel de Cervantes, last modified 2017, https://www.cervantesvirtual.com/nd/ark:/59851/bmctj0m8.
39. Moisés González Navarro, *La colonización en México, 1877–1910* (Mexico City: Talleres de Impresión de Estampillas y Valores, 1960), 1–10; Nancy Leys Stepan, *"The Hour of Eugenics": Race, Gender, and Nation in Latin America* (Ithaca, NY: Cornell University Press, 1991); Agustín F. Basave Benítez, *México mestizo: análisis del nacionalismo mexicano en torno a la mestizofilia de Andrés Molina Enríquez* (Mexico City: FCE, 1992); Alexandra Minna Stern, "From Mestizophilia to Biotypology: Racialization and Science in Mexico, 1920–1960," in *Race and Nation in Modern Latin America*, ed. Nancy P. Appelbaum et al. (Chapel Hill: University

Porfirian elites identified two main conditions that would contribute to the improvement of the "Indigenous race": (positivist) education and immigration. Many thought that the former should be made free and accessible to the general population so that it learned the principles for contributing to the improvement of the nation. The latter implied that by bringing in the supposedly most productive of "races," the new arrivals could make "vacant lands" work,[40] teach "Indians" by becoming an example for them to follow, and, in the case of Europeans, "mix" with the locals to accelerate the Mestizaje process. Justo Sierra, a prominent member of the generation of conservative-liberals, became one of the main advocates of both measures. From his congressional post, he launched an initiative for a constitutional amendment that declared primary school to be state-funded, obligatory, and free for all Mexicans,[41] and he backed the creation of a public National University founded on scientific principles.[42] With regard to immigration, Sierra, like many of his contemporaries, advocated for the arrival of white Europeans "so as to obtain a cross with the indigenous race, for only European blood can keep the level of civilization . . . from sinking, which would mean regression, not evolution."[43] The government supported this by offering colonization subsidies. From 1880, Secretary of Development Carlos Pacheco asked foreign legations to promote trade and invite investors interested in colonization and industrial projects, with a preference for white settlers.[44] As seen in his memoir, Bulnes also supported the idea of the superiority of Europeans and, therefore, the need to encourage their immigration to Mexico. He reinforced this in his book *The Future of the Latin American Nations*, where he suggested dividing humanity into three racial groups, each based on dietary traditions associated with the consumption of wheat, corn, or rice. Supposedly basing his ideas on recent scientific studies, he argued that the wheat diet was richer in nutritional value, which explained the moral and intellectual superiority of Europeans.[45] Throughout the Porfiriato, intellectuals continued

---

of North Carolina Press, 2003), 187–209; Miguel Ángel Avilés Galán, "Measuring Skulls: Race and Science in Vicente Riva Palacio's México a través de los siglos," *Bulletin of Latin American Research* 29, no. 1 (January 2010): 85–102.

40. So-called vacant lands were often inhabited by First Nations.
41. The idea was not new, as since the 1867 educational reform discussed in the last chapter, the law provided for state-controlled, publicly funded, free primary education in Mexico City for those who had no money. Yet the measure had been poorly enforced and was only accessible in the capital. Sierra's proposal, launched in the early 1880s, became law in 1888. Hale, *Transformation of Liberalism*, 225–31.
42. He launched this idea in 1881, but it only became a reality in 1910. Justo Sierra, *Evolución política del pueblo mexicano* (Caracas: Biblioteca Ayacucho, 1985), xv. This work, originally published in 1902, synthesizes Sierra's view on the historical evolution of the Mexican peoples in a similar vein to that of *México a través de los siglos*.
43. Cited in Raymond B. Craib III, "Chinese Immigrants in Porfirian Mexico: A Preliminary Study of Settlement, Economic Activity and Anti-Chinese Sentiment," *Research Paper Series* 28 (May 1996): 4.
44. Jason Oliver, *Chino: Anti-Chinese Racism in Mexico, 1880–1940* (Champaign: University of Illinois Press, 2017), 62.
45. Francisco Bulnes, *El porvenir de las naciones latinoamericanas ante las recientes conquistas de Europa y Norteamérica. Estructura y evolución de un continente* (Mexico City: El pensamiento vivo de América, 1899).

discussions that had been ongoing starting with Juárez's governments on the best "races" that should immigrate to Mexico.⁴⁶

Chinese in Mexico: Debates and Arrivals in Early Porfirian Governments

One of the first and fiercest public advocates for Chinese immigration to Mexico among the liberal ranks was Matías Romero Avendaño. From 1855 until his death in 1898, Romero held different posts in the successive administrations of Juárez, Lerdo, Díaz, and González, mostly within the Ministry of Foreign Affairs, specifically as the country's representative in Washington.⁴⁷ In 1875, while working as a coffee producer in the southern state of Chiapas, he published his first article in favor of Chinese immigration.⁴⁸ Speaking from his experience trying to grow export crops, he concluded that "where we most urgently need immigrants is on our coasts, both because they are the least populated and because they produce the agricultural products which bring the best prices abroad. . . . Also because being so close to the sea it is easier to export without paying high charter fees." Since both Europeans and Indigenous people refused to work under harsh coastal conditions,⁴⁹ "the only colonists who could establish themselves or work there are Asians, coming from climates similar to ours, primarily China."⁵⁰ In 1876 he published a follow-up article suggesting that increased trade relations with China and Japan would benefit the national economy.⁵¹

Romero's arrival in Washington in 1882 coincided with the implementation of the Chinese Exclusion Act, which barred legal immigration of Chinese male laborers

---

46. On this subject, see González Navarro, *La colonización en México*; and Moisés González Navarro, *Los extranjeros en México y los mexicanos en el extranjero, 1821–1970* (Mexico City: El Colegio de México, 1993).
47. In 1855, at age eighteen, he began working in the Ministry of Foreign Affairs. His most recurrent posts in the different governments were Finance Secretary, from January 1868 to June 1872, from May 1877 to April 1879, and from March to December 1892, as well as Mexican representative in Washington in the periods 1859–1867, 1882–1892, and 1893–1898. "Matías Romero (1837–1898)," in *Instituto Matías Romero. XXV Aniversario* (Mexico City: SRE, 1999), 107–40; Sergio Silva Castañeda and Graciela Márquez Colín, *Matías Romero and the Craft of Diplomacy: 1837–1898* (Mexico City: SRE, Instituto Matías Romero, 2018), 45, 57.
48. To learn about Romero's businesses in Chiapas and how they related to his diplomatic performance, see Mónica Toussaint Ribot, "Los negocios de un diplomático: Matías Romero en Chiapas," *Latinoamérica. Revista de estudios Latinoamericanos* 55 (2012): 129–57.
49. In 1877, the then Development Secretary Vicente Riva Palacio, in his report to the Congress, agreed with Romero on this point and recognized that "the government's lack of funds and the nation's lack of a transportation network" hampered European immigration to Mexico. "Europeans would not accept the lifestyle of the Mexican laborer [. . . and] wished to settle near population centers, where they are not needed." Cited in Kenneth Cott, "Mexican Diplomacy and the Chinese Issue, 1876–1910," *The Hispanic American Historical Review* 67, no. 1 (February 1987): 64.
50. Matías Romero, "Inmigración china en México," Revista Universal (August 1875). Reproduced in Josefina MacGregor, *Matías Romero. Textos escogidos* (Mexico City: CNCA, 1992), 472–73. On page 474, he talks about a "racial affinity" between the Mexican Indian and the Chinese.
51. The second article was published in *El Correo del Comercio*. Chao, *Chinese in Mexico*, 26. See also Grace Peña Delgado, *Making the Chinese Mexican: Global Migration, Localism, and Exclusion in the U.S.–Mexico Borderlands* (Stanford, CA: Stanford University Press, 2012), 13–15.

to the United States initially for a period of ten years.⁵² Just as the so-called coolie trade that brought hundreds of thousands of Chinese laborers to the Americas, often under harsh conditions, was coming to an end,⁵³ a new form of prohibition targeting them emerged, that of the "illegal alien."⁵⁴ For Romero, who interpreted the "coolie" experience in a positive (and positivist) light, the US Act represented an opportunity for Mexico. In Romero's eyes, the Chinese possessed advanced agricultural skills, adapted to harsh nature and labor conditions, accepted low wages, and worked hard. These attributes had helped many regions develop economically. Now that the US was expelling them, they were essentially a captive market.

For their own reasons, Chinese themselves increasingly began to see Mexico as a viable destination.⁵⁵ The exclusion laws, first in the United States and later in British Columbia, Canada,⁵⁶ coincided with the Mexican government's efforts to attract laborers for various projects in railroads, mining, and agriculture. Many arrived directly from Canton or from neighboring countries to Mexico, such as the US, Canada, or Cuba. While a large number of them did take jobs as unskilled workers in rural and urban settings throughout the country, particularly in the north, many ended up transitioning to retail and became owners of small businesses such as restaurants, laundries, and grocery stores. They were often supported by family and friends who had already established successful businesses in neighboring US states, forming a transnational commercial orbit that benefited all those involved. In addition, Chinese used Mexico as a trampoline to access the United States, taking advantage of the lack of surveillance over large parts of the border.⁵⁷ It is estimated that at least 17,000 Chinese entered the United States via Mexico and Canada in the decades following the Exclusion Act.⁵⁸

Within this context and in his key post in Washington, Matías Romero found himself pressured by opposing sides to either prevent or encourage Chinese arrivals

---

52. "Chinese Exclusion Act (1882)," National Archives, accessed October 5, 2023, https://www.archives.gov/milestone-documents/chinese-exclusion-act#:~:text=It%20was%20the%20first%20significant,immigrating%20to%20the%20United%20States. The restrictions on Chinese immigration to the United States were extended in 1892 and 1904 and it was not until 1943 that the bar was entirely lifted. Chao, *Chinese in Mexico*, 1.
53. Between 1847 and 1874, at least 1.5 million Chinese left from the southeastern coast of China for Southeast Asia, Australia, and the Americas. In Latin America, so-called Chinese coolies or contract laborers went mainly to Cuba (around 150,000) and Peru (around 100,000). Most of them left from the less regulated port of Macao, administered by the Portuguese. Elliott Young, *Alien Nation: Chinese Migration in the Americas from the Coolie Era through World War II* (Chapel Hill: University of North Carolina Press, 2014), 33.
54. As Elliot Young explains, "no federal bureaucracy had the power to exclude a particular ethnic or national group until the Chinese exclusion act of 1882. Thus, the Chinese became the first group not only to bear the stigma of aliens but to be enmeshed in a federal bureaucratic and legal system that produced them as 'illegal' as well." Young, *Alien Nation*, 12, 93.
55. Oliver, *Chino*, 63.
56. To understand the path that the United States followed to get to the 1882 Exclusion Act, see Young, *Alien Nation*, 101–14. As for Canada's 1885 Act to Regulate and Restrict Chinese Immigration, which initially levied a $50 head tax on Chinese immigrants and limited the number of Chinese passengers on each ship, see pages 104–6.
57. Chao, *Chinese in Mexico*, 4–5, 97. The book is full of interesting examples that support these conclusions.
58. Young, *Alien Nation*, 160.

to Mexico. On the restrictive side, US officials complained of Mexico's free immigration policy, which allowed Chinese to easily land in Mexico and pass unchecked into their country.[59] To this, Romero often cited the eleventh article of the Mexican Constitution of 1857, which stated: "All men have the right to entering and leaving the Republic, of traveling through its territory, and of changing their residence without the necessity of letters of security, passports, or other similar requisite."[60] Domestically, many still argued along the lines of the renowned Mexican historian, geographer, and jurist Manuel Orozco y Berra: "opening the doors . . . to the leftover elements, expelled from other nations, . . . [would imply] receiving the least enlightened people, and receive them as they are: with all their vices, with all their ignorance, with all their inconveniences. The determining reasons for the Chinese expulsion [from the U.S.] . . . are well-known: their dominating vices, their shrewd, indolent and distrustful character, their repulsion to good, their hate to everything that is not from their country, their sordid greed, and many other faults that might be even unknown to our race."[61] The Mexican consuls in San Diego and San Francisco, who had closely followed the Exclusion Act debates in California, where the anti-Chinese rhetoric was the most virulent, were also unsupportive of bringing Chinese nationals to Mexico.[62]

At the same time, Romero found allies within the government, even though most still held strong prejudices. For instance, Manuel Zapata Vera, author of an 1882 report on Chinese experiences on the American continent, saw them as a "necessary evil." Despite their supposedly multiple vices, Zapata acknowledged that Mexico needed them to make tropical lands productive. But in order to avoid "the problems [of the] nations that opened their doors [to Chinese] indiscriminately," he proposed a "discrete and prudent system" consisting of two simple steps: "1) choose the more adequate zones, that is, those that need more foreign hands, and 2) at first limit the numbers of immigrants and . . . examine the results, without prejudices and with the purpose of taking advantage of this new element if it proves to be useful for the development of our agriculture."[63] That same year, Ignacio Mariscal, then representative in Great Britain, sent a letter to Romero stating that Mexico City's politicians did not see much problem with the increasing arrivals of Chinese as they would be channeled to colonization projects in uninhabited areas. Even if they decided to move to populated areas, they would either eventually return to their homeland or not mix with locals, who he believed were uninterested in Chinese.[64] Convinced that the only way to increase their presence in Mexico was via diplomatic ties, Romero initiated the first informal exchanges with Cheng Tsao Ju,

---

59. Young, *Alien Nation*, 99.
60. Cited in Young, *Alien Nation*, 100.
61. AHGE-SRE, AEMEUA, Leg. 67, exp. 22, 1879–1881, 51–59.
62. See documents in AHGE-SRE, 15-2-69.
63. Parts of the report are reproduced in Manuel Zapata Vera, "Chinese Immigration to Mexico," *El Economista Mexicano* 1 (February–July 1886): 139–40.
64. Oliver, *Chino*, 65.

his Chinese peer.[65] The *Mount Lebanon* case would give him the push he needed to formally pursue this objective.[66]

The Controversial Ship

As mentioned earlier, modernization of the transport infrastructure became key for the Porfirian economic model, which relied on export commodities and often required the import or transfer of labor. Throughout his presidency, Díaz devoted the largest proportion of public investment to the construction of railroads,[67] expanding the network from 660 kilometers of track at the beginning of his first presidency to close to 20,000 kilometers by the time he left office in 1911,[68] making it, after Argentina, Latin America's second largest national railway system.[69] Like the rest of the world, with the exception of England, Mexico was unable to commence a railroad system with its own resources.[70] After a series of concessions to its state governments failed,[71] Díaz's administration resorted to foreign capital. It first granted subsidies in cash, paying an average of 9,500 pesos per kilometer built.[72] Other strategies included land grants, the exemption from import duties on construction materials, and the hiring of foreign contractors.[73] By the time the *Mount Lebanon* controversy erupted in 1884, Mexico had just inaugurated its second railroad line, linking Mexico City with Ciudad Juárez, on the border with Texas—built with US capital. The country now claimed 4,658 kilometers of new track.[74]

---

65. AHGE-SRE, AEMEUA, 1879–1890, 1881–1883, 1888, Leg. 123, exp. 1.
66. For an extended discussion on the wide array of views concerning the immigration of Chinese to Mexico during the Porfiriato, see González, *Extranjeros en México*, 163–78; Cott, "Mexican Diplomacy," 63–85. See also Robert H. Duncan, "The Chinese and the Economic Development of Northern Baja California, 1889–1929," *The Hispanic American Historical Review* 74, no. 4 (November 1994): 615–47.
67. Priscilla Connolly, *El contratista de Don Porfirio. Obras públicas, deuda y desarrollo desigual* (Mexico City: ColMich/UAM/FCE, 1997), 82.
68. Connolly, *Contratista*, 82 and Sandra Kuntz Ficker, "Fuentes para el estudio de los ferrocarriles durante el Porfiriato," *América Latina en la historia económica. Boletín de fuentes* 13–14 (Jan.–Dec. 2000): 137.
69. Paolo Riguzzi, "Los caminos del atraso: tecnología, instituciones e inversión en los ferrocarriles mexicanos, 1850–1900," in *Ferrocarriles y vida económica en México, 1850–1950. Del surgimiento tardío al decaimiento precoz*, ed. Sandra Kuntz and Paolo Riguzzi (Mexico: El Colegio Mexiquense/UAM-X/FNM, 1996), 31.
70. Paolo Riguzzi, "Propiedad, propietarios y recursos nacionales en los ferrocarriles mexicanos, 1870–1905," in *Memorias del Tercer Encuentro de Investigadores del Ferrocarril* (Puebla: Museo Nacional de los Ferrocarriles, 1996), 211.
71. The Mexican government granted its first railroad concession in 1837. Between 1850 and 1876, it supported some fifty more; however, 93 percent of them produced no results. The first line, linking Mexico City with Veracruz, was finally completed in 1873. Riguzzi, "Caminos," 55. To understand the reasons behind these failures, see pages 31–97.
72. If one excludes the large subsidy of 44 million for the 309 kilometers of the Tehuantepec railroad, then the average reduces to 6,500 pesos per kilometer built. Riguzzi, "Caminos," 74.
73. Connolly, *Contratista*, 81–86. According to Sandra Kuntz, the Mexican railroads were composed of English tracks, European and US fuel, and US wagons and locomotives. Sandra Kuntz, "Los ferrocarriles y la formación del espacio económico en México, 1880–1910," in *Ferrocarriles y obras públicas*, ed. Sandra Kuntz and Priscilla Connolly (Mexico City: Instituto Mora/ColMich/Colegio de México/IIH-UNAM, 1999), 112.
74. Riguzzi, "Caminos," 64; Kuntz, "Ferrocarriles," 106. See also David M. Pletcher, *Rails, Mines, and Progress: Seven American Promoters in Mexico, 1867–1911* (Port Washington: Kennikat, 1972); Arturo Grunstein,

In the case of steamships, in the 1880s the federal government began financing Mexican firms to create a national merchant navy moved by steam. In 1882 it granted its first two concessions to Mexican companies traveling abroad in the Gulf of Mexico: the Trasatlántica Mexicana and the Mexicana Continental. While up to that point the government had generally accorded foreign companies around 2,000 pesos per round trip, the former was to receive ten times more for its transatlantic lines. The latter maintained the usual 2,000-peso subsidy for its New York–Havana–Progreso–Veracruz route. In the Pacific, the first concession was for CMNP.[75] The contract, signed in March 1884 between the federal government and businessmen Salvador Malo, Guillermo Vogel, and Luis Larraza, established that CMNP would have two lines. The first would link various Mexican ports from Guaymas in the north to Soconusco in the south, with the possibility of extending it to the United States and Central America to cover the entire San Francisco–Panama route. There was no subvention granted to this line, only tax exemptions. The second would travel from a Mexican Pacific port—any of those between San Blas and Acapulco—to Yokohama, Hong Kong, and Manila. Besides tax exemptions, the government would reimburse 19,000 pesos per round trip—or 18,000 if the vessel skipped Manila. The company would also receive sixty-five pesos per European immigrant and thirty-five pesos per Asian laborer over seven years of age. Women could not exceed half of the total migrants brought. Each vessel was entitled to bring up to 1,000 passengers and the company would be fined if it did not bring at least 1,200 laborers every six months. Operations had to start within fifteen months of signing.[76] While the contract did not specify the place where the laborers would work, the first destination for most would be the Tehuantepec railroad, of which Salvador Malo was a subcontractor.[77] While there was no specification in the contract as to what CMNP's vessels were supposed to take to Asia in return for Chinese laborers, silver was thought to be the main exchange commodity. In this regard, the senior development official Manuel Fernández Leal, one of the members of the Mexican Astronomic Commission discussed in Chapter 2,[78] together with the Foreign Secretary, said that inaugurating maritime and commercial relations with

---

Railroads and Sovereignty: Policy-Making in Porfirian Mexico (Los Angeles: University of California Press, 1994).
75. Roberto García Benavides, *Hitos de las comunicaciones y los transportes en la historia de México* (Mexico City: Secretaría de Comunicaciones y Transportes, 1988), 282–84.
76. AHGE-SRE, 44-6-35, 102–4.
77. AHGE-SRE, L-E-1515, 3. The Tehuantepec railroad is discussed further in Chapter 5.
78. In relation to silver, the Commission's head, Francisco Díaz Covarrubias, had noted in his memoir that "the Mexican coin . . . still conserves so much prestige, that not even the Anglo-American trade dollar has been able to destroy it." But without a direct route from Mexico to Asia, the coin traveled "the longest route, covering a trajectory of more than seven thousand leagues, and leaving in its wake part of its value; meanwhile, if it were sent by us to Asia, it would not have to travel more than two thousand leagues, nor would it lose its value as it does today by powering European commerce." Francisco Díaz Covarrubias, *Viaje de la Comisión Astronómica Mexicana al Japón para observar el tránsito del planeta Venus por el disco del sol el 8 de diciembre de 1874* (Mexico City: Políglota, 1876), 143. Translation by the author and Martin McLennan.

China and Japan was beneficial because Mexicans "could send pesos ... before the governments of England, the United States and other empires managed to make other coins circulate." This was particularly important "now that the [Mexican] coin value has diminished so much in European markets."[79] Regardless of the practicability of this measure—after all, the value of the Mexican coin was then dictated by European stock markets—the discursive weight of silver as an exchange commodity with Asia continued to play an important role in transpacific voyages in the 1880s. It is worth mentioning that, if successful, this line would be in direct competition with the Pacific Mail Steamship Company (PMSS) as well as with various British and French steamers running between Yokohama and Hong Kong.

CMNP's owners inaugurated their transpacific service with the chartered steamer *Mount Lebanon*. Chartered steamers or tramps across the Pacific became quite common in the 1880s, most of them being of British, German, or Norwegian manufacture.[80] The British-made *Mount Lebanon* landed on October 3, 1884, at dock 4c of the Hong Kong harbor.[81] When the shipmaster asked for a passengers' license to embark Chinese laborers for Mexico, the governor of Hong Kong, Lord Ferguson, immediately refused. Since 1855, the British authorities had enacted a series of laws regulating the hugely chaotic, abusive, and highly profitable traffic of Chinese laborers within and outside their territories. While mistreatment often continued, the laws served as a foundational framework that migrants, merchants, and authorities could turn to in order to resolve disputes.[82] Lord Ferguson argued that, based on these laws, he could not allow Chinese subjects to travel as slaves to a country—Mexico—that had no relations with China or England.[83]

The British authorities became more flexible after the Principal Secretary of State for the Colonies, the Earl of Derby, received a letter from Theodor Schneider, a representative of the British firm Jardine, Matheson & Co., in his London office. It stated that CMNP had requested the services of his firm to manage its business in Hong Kong. As the company's representative for the case, Schneider defended CMNP's interests using an argument that British diplomats would not easily dismiss: "I may mention that the trade will be carried on in British-owned vessels [from Smith & Service] & under the British Flag, and the opinion of the principal merchants is that the establishment of the line in question will largely benefit the

---

79. AHGE-SRE, 44-6-47, 1–2. Fernández Leal published a short study on the topic of silver in 1886 titled "Producción de oro y plata en las principales naciones del mundo." Rodrigo Antonio Vega y Ortega Baez and Gustavo Enrique Flores Herrera, "Un funcionario experto. La producción geográfica del ingeniero Manuel Fernández Leal, 1877–1911," *Investigaciones Geográficas* 109 (December 2022), https://www.scielo.org.mx/scielo.php?pid=S0188-46112022000300102&script=sci_arttext#B21.
80. E. Mowbray Tate, *Transpacific Steam: The Story of Steam Navigation from the Pacific Coast of North America to the Far East and the Antipodes, 1867–1941* (New York: Cornwall Books, 1986), 86.
81. *The China Mail*, October 4, 1884, 2, 4.
82. For a study on the evolution of the laws as well as the abuses of Chinese laborers in the nineteenth century, see Persia Crawford Campbell, *Chinese Coolie Emigration to Countries within the British Empire* (London: BiblioLife, 2009), 86–160.
83. AHGE-SRE, 44-6-35, 1.

Colony of Victoria, as it has been proved in all similar cases that a large trade has been the result of such emigration."[84]

At the request of CMNP's owners, the Mexican authorities also responded to Lord Ferguson's refusal. On October 8, Ignacio Mariscal—the Mexican envoy to London who was negotiating the renewal of relations with Britain—explained personally to Lord Edmond Fitsmaurice, the British Under-Secretary of State for Foreign Affairs, that his country's liberal laws forbade slavery. Additionally, according to the contract, all Chinese laborers would travel by their own free will and would be informed in their own language of the conditions of the agreement. Throughout the crossing, the company would provide all passengers with clean lodging and sufficient food. In order to supervise this, the government would appoint at least one inspector per trip.[85] Lord Edmond Fitsmaurice agreed to submit the Mexican plea to the Earl of Derby.[86] On October 31, the British replied that they would request authorization from the Chinese government to send laborers to Mexico.[87]

For the Chinese authorities, *Mount Lebanon* and its troubles simply were not a priority. Faced with a war with France over the control of Tonkin, the Emperor did not have the time to deal with a Mexican ship,[88] nor did he want to compromise the security of his subjects, especially after the mistreatment they had been receiving in various countries where they had migrated. The fact that Mexico and China had no bilateral relations made things more complicated as it would be even more difficult to defend the laborers' interests. Marquis Tseng, the Chinese representative in London, explained this to Mariscal during an early November meeting. Tseng therefore suggested that *Mount Lebanon* load whatever it could and then leave, for the official answer from the Chinese authorities was going to take some time.[89]

In Washington, Matías Romero approached his Chinese counterpart, Cheng Tsao Ju, not only over the issue of *Mount Lebanon*, but also to discuss a possible treaty between the two countries, as this was the only long-term solution he envisioned to avoid these kinds of problems. Cheng replied that he had enough complaints to deal with from the overseas Chinese—he was assigned all subjects living in the United States, Peru, and Cuba—and was not in a position to focus on this matter.[90]

---

84. The National Archives, Foreign Office, FO50/451, 4–5.
85. AHGE-SRE, 44-6-35, 103.
86. AHGE-SRE, 44-6-35, 9.
87. AHGE-SRE, 44-6-35, 21.
88. Since the beginning of the decade, China and France had engaged in military confrontations for the control of Tonkin (present-day northern Vietnam). During the fall of 1884, several battles were taking place and the Chinese were facing considerable losses. On this war and Chinese international relations at the time, see L. M. Chere, *The Diplomacy of the Sino-French War (1883–1885): Global Complications of an Undeclared War* (Notre Dame: Cross Cultural Publications, 1988); B. Elleman, *Modern Chinese Warfare, 1795–1989* (London: Routledge, 2001).
89. AHGE-SRE, 44-6-35, 27–32.
90. These were the places with the largest Chinese populations in the Americas and with which China had bilateral relations (Cuba was then still a colony of Spain). Since China had no permanent representative head in Cuba or Peru, Mr. Cheng often had to travel and stay there for months. See file L-E-1983 in AHGE-SRE to see the exchanges between Romero and Cheng Tsao Ju as well as the following bibliography on the Chinese

In his words, as transcribed by Romero, "the Emperor's government believed that, since there was no traffic nor important official relations between both countries, and not being easy for him to accredit a minister in Mexico, it would be preferable to defer the celebration of a treaty for later, when the increase in trade and relations made it necessary."⁹¹ In November, Cheng left for Peru and talks stalled. Once he had returned, the minister fell sick and left his post. His successor, Chang Yen Huan, also prioritized his compatriots' complaints throughout his territory. As a consequence, talks between him and Romero were repeatedly interrupted.⁹²

Meanwhile, the British representative in Peking, H. Parker, received a telegram from London urging him "to press for the consent of the Yamen at all events for this [*Mount Lebanon*] voyage." He immediately presented his petition personally to the Prince and his ministers, highlighting that "great expense was incurred by delay" in CMNP's case. "I pointed out that as the conditions under which this emigration was conducted were declared to be satisfactory by Her Majesty's Government, and as the immigrants when in Mexico would receive British protection, the Chinese government might feel confidently assured that they [the Chinese] would be well treated."⁹³ He also "thought it not out of place . . . to remind His Highness and Their Excellencies that Your Lordship's desire to obtain their consent was an evidence of good feeling and friendly consideration, as it was within the power of Her Majesty's government to sanction emigration from Hong Kong to any part of the world."⁹⁴ After much hesitation, the Chinese approved the emigration of Chinese aboard *Mount Lebanon* but confined it to a single voyage, adding that in the future prohibition would be renewed. They remarked that this was "given in deference to Your Lordship's special request and to their regard for the existing friendly relations with Great Britain."⁹⁵ The Yamen then telegraphed its response in early December to Foreign Secretary José Fernández in Mexico City. It stated that the Emperor did not approve the emigration to Mexico because of the lack of a bilateral treaty. Yet, considering that *Mount Lebanon* was already in Hong Kong, the company could embark Chinese laborers on this occasion only, provided they were accorded British protection in the Mexican territory.⁹⁶

---

contract laborers in the Americas: Mauro García Triana and Pedro Eng Herrera, *The Chinese in Cuba, 1847–Now* (Lanham, MD: Lexington Books, 2009); Evelyn Hu-DeHart, "On Coolies and Shopkeepers: The Chinese as Huagong (Laborers) and Huashang (Merchants) in Latin America/Caribbean," in *Displacements and Diasporas: Asians in the Americas*, ed. Wanni W. Anderson and Robert G. Lee (New Brunswick, NJ: Rutgers University Press, 2005); Adam McKeown, *Chinese Migrant Networks and Cultural Change: Peru, Chicago, Hawaii, 1900–1936* (Chicago: University of Chicago Press, 2001).

91. Cited in Vera Valdés Lakowsky, *Vinculaciones sino-mexicanas: albores y testimonios (1874–1899)* (Mexico City: UNAM, 1981), 99.
92. Valdés, *Vinculaciones*, 100–104.
93. The National Archives, Foreign Office, FO50/451, 74–75. The Yamen was the English term used in the late Qing dynasty to refer to the Chinese office of Foreign Affairs.
94. The National Archives, Foreign Office, FO50/451, 75–76.
95. The National Archives, Foreign Office, FO50/451, 76–77.
96. AHGE-SRE, 44-6-35, 23.

The Mexican government refused the requested British protection for the immigrants: accepting another country's jurisdiction inside Mexican territory was out of the question. After so many foreign military interventions in Mexico and the long tradition of European expansionism, Porfirian diplomacy drew a red line when it came to compromising the country's political sovereignty.[97] Even though Porfirian diplomats were eager to resume ties with England and end this controversy, on principle they could not compromise the right to treat foreigners in accordance with Mexican legislation, as it would set a dangerous precedent. Mariscal had been negotiating the resumption of ties with London under the premise of a reciprocal renunciation of previous claims. This meant that Mexico would not demand any compensation related to the English support of Maximilian's French invasion in exchange for the British withdrawing its defense of the so-called London bondholders, a group of British investors that claimed a debt of over £10 million dating back to the 1820s.[98] This also included an understanding that both countries were to consider each other's legislation on equal and reciprocal terms. According to this reasoning, the Chinese did not need British protection as they would be treated according to Mexican laws. Mariscal also privately questioned the alleged concern for Chinese wellbeing and saw the Hong Kong governor's refusal as "an excuse... as with its monopolizing tendencies, the British government must not like the establishment of a direct traffic that could flourish [... between Mexican] and Asian ports."[99] CMNP's owners would eventually share this mistrust as, according to them, the "Hong Kong governor allows the Chinese emigration to various other countries, even though they have no [bilateral] treaty."[100] On December 14, at the insistence of CMNP's owners, Mariscal ended up accepting the Chinese proposal, but under two conditions: the British would provide good offices instead of full protection and this would have a one-year limit.[101] The Chinese responded that they wanted full protection rather than merely good offices and thought that one year was too short, "considering the distance and the time taken in all negotiations."[102]

In January 1885, Porfirio Díaz reclaimed the presidency and made the resumption of ties with England a priority.[103] Upon taking office, Díaz immediately summoned Mariscal to Mexico City. After over a year and a half in Europe negotiating with the British authorities, Mariscal took a transatlantic steamship and sailed back to Veracruz, then took the train to Mexico City. On January 19, he became the

---

97. Jorge A. Schiavon et al., eds., *En busca de una nación soberana. Relaciones internacionales de México, siglos XIX y XX* (Mexico City: SRE/CIDE, 2006), 15, 25, 152.
98. Silvestre Villegas Revueltas, "La deuda inglesa: el componente de la relación anglo-mexicana," in *En busca de una nación*, 197. See the entire article to understand the evolution of the negotiations between Mexican and British diplomats to re-establish bilateral relations.
99. AHGE-SRE, 44-6-35, 21.
100. AHGE-SRE, 44-6-35, document from April 4, 1888 (no page number).
101. AHGE-SRE, 44-6-35, 39.
102. AHGE-SRE, 44-6-35, 41.
103. Díaz's first presidential term lasted from 1876 to 1880. In December 1884, he reassumed the presidency for a second term.

Secretary of Foreign Affairs, a post he would hold until his death in April 1910.[104] With his ample knowledge on the subject and with the process already quite advanced, he was soon able to re-establish official ties with London. In order not to compromise this achievement, the *Mount Lebanon* case was put aside and negotiations did not advance for months. In April, after a desperate plea from Malo, Vogel, and Larraza, Mariscal finally agreed to take out the one-year limit but refused to change the term of good offices.[105] Yet he guaranteed that the term would imply full protection for the immigrants. As a consequence, on May 1, 1885, Lionel Carden, the representative of Great Britain in Mexico, wrote a favorable note to Mariscal: "I have the honor to inform your Excellency that today I have received a telegram from Earl Granville, informing me that the Chinese have agreed to the emigration of their subjects to Mexico, but only for one voyage, and that the Governor of Hong Kong has been instructed accordingly."[106] By then, *Mount Lebanon* had already left for Saigon. Its British owners, Smith & Service, had assigned it to a more profitable business starting in March. Even though it returned to Hong Kong in May, it was not assigned to a transpacific trip but rather was sent back to Saigon.[107]

At this point, precisely when negotiations seemed to have neared an end, Smith & Service introduced a new element into the dispute. In a letter to Whitehall, the British firm complained that CMNP had refused to pay its dues: "there have been various communications from the London Brokers of the Mexicana Company [Jardine, Matheson & Co.] but neither any payments of our claim nor proposal to pay even a part of it. In this morning's letter they threaten to charter another steamer for the Emigration permitted by your Lordship, leaving 'Mount Lebanon' out of the business altogether. . . . [Will] the Mexican company . . . be permitted to send Emigrants by another steamer, after having kept 'Mount Lebanon' without any remuneration for over six months?"[108] Even though the contract signed by both parties established an advanced payment, Smith & Service claimed they had not received a penny from CMNP. Furthermore, the Mexican government stated that "the delay caused by Prohibition of Emigration by the British government [will] be considered 'Force Majeure' and the [CMNP] company exonerated from their contract."[109] Smith & Service therefore proposed that "your Lordship should still prohibit any Chinese emigration from Hong Kong [to Mexico], and ask the Chinese Government to issue a similar prohibition as to other Chinese ports . . . we who are British subjects have suffered a loss of not less than £8,000 sterling and . . . it does seem hard upon us to be relegated to the remedy of the very Laws of Mexico

---

104. Vera Valdés Lakowsky, "Ignacio Mariscal," in *Cancilleres de México, vol. 1* (Mexico: SRE, 1992), 579, 582–83.
105. AHGE-SRE, 44-6-35, 48, 50, 58.
106. AHGE-SRE, 44-6-35, 81.
107. See *The China Mail* from March 5 to May 15, 1885.
108. The National Archives, Foreign Office, FO50/451, 269.
109. The National Archives, Foreign Office, FO50/451, 187–88.

which your Lordship has evidently considered insufficient to protect the rights of Chinamen."[110]

Whitehall thus found itself in a dilemma: If it supported Smith & Service—a British firm, after all—then it would have to cancel the deal achieved with the Chinese and the Mexican governments and businessmen—who had employed another British firm, Jardine, Matheson & Co., to defend their interests—for a one-voyage solution, which had taken months to negotiate. To resolve the situation, it came up with the least damaging practical compromise: permission for the single voyage was extended specifically to *Mount Lebanon*.[111] With this decision Smith & Service's interests were protected, because it guaranteed that CMNP would have to use *Mount Lebanon*—and not another steamer—if it wanted to run one voyage from Hong Kong to Mexico. Simultaneously, the British could keep their word with Mexican and Chinese diplomats by maintaining the one-voyage deal, and thus support the other British company involved in the case, Jardine, Matheson & Co. The British diplomats did not request the Chinese authorities to forbid emigration to Mexico from other ports, as Smith & Service also demanded. After all, the Chinese were themselves against such emigration.

CMNP never recovered from the *Mount Lebanon* failure. On August 11, 1885, a Hong Kong officer informed Whitehall that "the Chinese Agent [hired by CMNP to recruit laborers for Mexico] has, for the present at least, abandoned the scheme and the emigration license granted to him has lapsed."[112] This seems to indicate that *Mount Lebanon* did not even complete the single voyage to which it was entitled. CMNP tried to restart operations several times up to 1888, but the British refused permission at every attempt, arguing that Mexico had no treaty with China and that it was therefore impossible for them to let Chinese laborers depart.[113] In November 1889, Salvador Malo signed a new contract with the Mexican government to create the Compañía Marítima Asiática Mexicana (CMAM), once again for the purpose of establishing direct transpacific routes and bringing laborers to Mexico. The terms of the contract were similar to those given to CMNP, and its fate would prove similar as well.[114] Malo sent a representative to Asia, who traveled to San Francisco and then sailed on to Yokohama on August 22 in one of PMSS's largest vessels, the 5,080-ton *City of Peking*.[115] He was unsuccessful in dealing with the Japanese but managed to start operations in Macao, a less regulated port administered by the Portuguese.[116]

---

110. Since CMNP had no offices in England, Smith & Service could not sue them in a British tribunal but rather had to do so in a Mexican one. The National Archives, Foreign Office, FO50/451, 283.
111. The National Archives, Foreign Office, FO50/451, 313, 315, 317.
112. The National Archives, Foreign Office, FO50/451, 319–20.
113. See the last documents in AHGE-SRE, 44-6-35 as well as *El Economista Mexicano* 9 (February–August 1890): 7.
114. AHGE-SRE, L-E-1515, vol. 4, 1.
115. See *El Tiempo*, September 6, 1890, and December 7, 1890, as well as Tate, *Transpacific Steam*, 28–29.
116. Enrique Cortés, *Relaciones entre México y Japón durante el Porfiriato* (Mexico City: SRE, 1980), 52–53. See AHGE-SRE, L-E-1515, vol. 4, 1–65 to learn about the exchanges between Malo and the Japanese authorities.

On September 29, 1890, the tramp *El Amigo* left Macao for the Mexican port of Salina Cruz with some 500 Chinese on board. They were destined to work on the construction of the remaining 180 kilometers needed to finish the Tehuantepec railroad.[117] The company claimed it had another steamship ready to sail on October 27,[118] but there are no records that show that this, or any other steamer from CMAM, ever crossed the ocean again. Salvador Malo died in 1901 and, with him, so too did attempts to create a Mexican transpacific steamship company.[119]

Conclusions: Mexico and Asia in the 1880s

In the 1880s, General Porfirio Díaz and the heterogeneous groups that supported him managed to create a relatively strong federal government that promoted a long-term modernization program for Mexico. Tailoring liberal and positivist notions towards their pragmatic needs, Porfirian circles legitimized their authoritarian and elitist practices as necessary means to achieve what they stated the country most urgently needed: material progress. Steamers played the vital role of transporting production to international markets and bringing back the commodities, laborers, and immigrants needed to accelerate its yield. Thus began a policy of subventions to create a local merchant navy to regulate and encourage international traffic. The backing of Mexican-owned lines on both the Atlantic and the Pacific was granted within years with analogous conditions and objectives: to create a pair of routes—one coastal and one transoceanic—on each seaboard.

According to the racialized and positivist opinions of many within Porfirian circles, the local populations, composed mainly of "unproductive Indians," were either insufficient, incapable, or unwilling to participate in the modernizing ventures the country supposedly needed. They therefore sought hardworking "white industrialist" immigrants, mostly from Europe, and mostly men, who would serve not only as economic forces but also as reproductive entities accelerating the "miscegenating" process that would eventually lead to the ideal "Mexican race," the Mestizo, formed by European and Indigenous blood, the pillar of a unified national identity.

Unable to attract the much desired whites, the idea to bring Chinese, who had proved industrious in other American countries as well as in the Spanish colony of Cuba, gained momentum despite polarized opinion. Those in favor alluded to their low cost and high productivity. Those opposed enumerated a long list of negative attributes supposedly inherent to their "race." Even those who defended the Chinese often failed to acknowledge their historical presence—indeed, since colonial times—or their role as "industrialists" and therefore reduced them to an

---

117. AHGE-SRE, L-E-1515, 21–29.
118. *El Tiempo*, October 14, 1890.
119. Cortés, *Relaciones*, 55.

alien, cheap, and productive labor force.[120] After much heated debate, the pragmatic approach of sponsoring limited and state-regulated Chinese immigration prevailed, resulting in the subvention of CMNP to inaugurate direct steam-powered maritime routes with Asia and bring Chinese laborers.

The Chinese themselves found Mexico to be an increasingly attractive destination. In the context of exclusion laws in the United States and western Canada, Mexico offered a relatively open immigration policy, job possibilities, and a promising geographical location close to family, friends, and compatriots who had settled in neighboring countries, as well as the potentiality to enter the United States through a porous border. Chinese presence thus increased. While a majority at first worked as laborers in agriculture, mining, and railway construction, many transitioned to become small business owners, often with the help of acquaintances who had previously settled and prospered, notably in the United States, forming mutually supportive transnational networks.

Efforts to create a national steam-powered merchant navy with links to transoceanic markets floundered on both coasts. In the Gulf of Mexico, the subventions granted in 1882 to the Trasatlántica Mexicana and the Mexicana Continental proved insufficient. The first company declared bankruptcy in 1886 and the second a few years later.[121] On the west coast, CMNP failed to successfully navigate British diplomacy, notably in Hong Kong, where it controlled the transit of people, particularly of Chinese laborers. Additionally, its businessmen dominated the industry of chartered steamers, which the Mexicans had no option but to resort to since they could not fabricate boats at home.

CMNP's case proved to be ambivalent for Mexican diplomats. On the one hand, those who supported the transoceanic venture failed to convince the English and Mexican authorities to help them overcome the legal obstacles on both sides of the Pacific. On the other hand, Porfirian diplomats did manage to re-establish ties with Great Britain and defend their key interests, particularly in relation to impeding foreign jurisdiction inside Mexican territory. Matías Romero's formal petition to his Washington counterpart in relation to CMNP marked the beginning of a long process of initiating bilateral relations with an Asian country.[122]

Even though the company's regular transpacific runs never materialized, the case study remains useful as it portrays steamships—even those that failed to sail—as in-between sites where different ideologies, interests, and points of view from both sides of the Pacific, and even beyond, coincided and collided. In addition, while macro-histories often present national objectives and actors as unitary and fixed, portraying the nineteenth-century Pacific as an unchallenged "British

---

120. Elliott Young notes that in the case of Cuba, Chinese remained outside the foundational myths of national unification despite ancestral connections. A similar conclusion can be reached for Mexico. Young, *Alien Nation*, 14.
121. "Nuestra Marina Mercante," *El Economista Mexicano* 10 (August 1890–January 1891): 43; García, *Hitos*, 283.
122. Chapters 4 and 5 will explain the culmination of this process with Japan and China respectively.

lake," CMNP's case proves this false. For instance, British firms and bureaucrats in London and Peking first collaborated with Mexican businessmen to initiate transpacific runs, in opposition to their authorities in Hong Kong. Throughout the negotiations, alliances and oppositions evolved, as was also the case among Porfirian circles. As for Chinese diplomats, facing European imperialist threats within their territory, they opted to concentrate their efforts at home. Recent mistreatment of their subjects made them hesitant to send laborers to a country with no official reciprocity. Their foreign diplomats were not interested either as they were too busy supporting those Chinese who had migrated earlier. Romero's counterpart, Cheng Tsao Ju, had made it clear: bilateral relations with Mexico were simply not a priority. This contrasted substantially with the Japanese posture. As we will see in the next chapter, the Japanese authorities saw Mexico not as a distractor, but as an opportunity in their quest to gain more equitable treatment from imperial powers.

4
Gaelic, 1897
The Japanese Colonization Project in Mexico

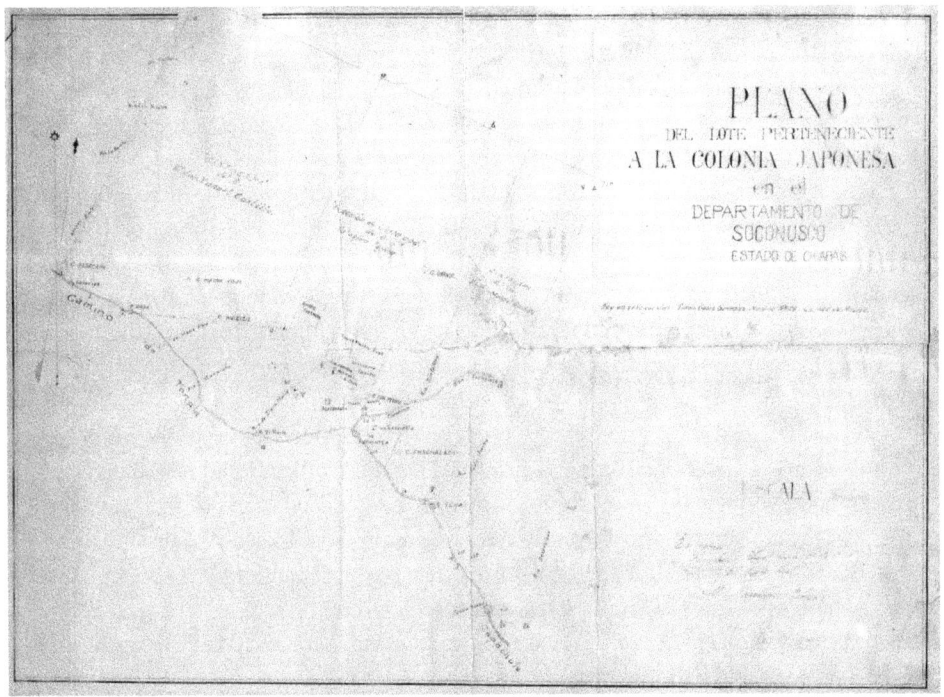

Map 4
M. Jaimes, *Plano del lote que pertenece a la colonia Japonesa en el Soconusco* [Map of the land belonging to the Japanese colony in Soconusco]. Map. Chiapas: the twentieth century. From Mapoteca "Manuel Orozco y Berra," Chiapas 6. Accessed December 15, 2023. https://mapoteca.siap.gob.mx/cgf-chis-m2-v6-0499/.

On March 24, 1897, a distinctive group of thirty-four young Japanese men left Yokohama aboard the Occidental & Oriental Company's (O&O) *Gaelic*, a 4,000-ton British steamer. While at first glance nothing was out of the ordinary—after all, the transpacific crowd bound for San Francisco routinely included hundreds of Asian males traveling in steerage—these men were atypical in one sense: they were the first state-sponsored Japanese colonists headed for Latin America.[1]

Once in San Francisco, they transferred to a steamer on a milk run to Panama. On May 10, they disembarked in the port of San Benito, located less than fifty kilometers from the recently settled Mexican–Guatemalan border. They were greeted by Tono Kuraji and Enomoto Ryukichi, two advance envoys who had prepared for the group's establishment in Mexico. After a twelve-hour trek, they arrived to the city of Tapachula, where they spent three days buying provisions. They then walked for another three days. On May 16, they finally reached their destination, a 65,000-hectare land parcel in Escuintla, Chiapas.[2]

The entire episode had been problem-ridden since departure. First, the workers had to bargain for better contractual conditions for their colonization once aboard *Gaelic*. Furthermore, at sea, one member fell terminally ill. Once in Mexico it became apparent that the survivors knew little about their new land, and conditions were tougher than expected. Their destination was in the middle of a hot, humid jungle with no infrastructure. And while they were contracted to grow coffee for export, their seeds and tools would take weeks to arrive. To add insult to injury, the settlers lost their entire first food crop to scavenging animals. In short, the initial experience of the first Japanese *colonos* (colonists) in Mexico was nothing short of a disaster.[3]

While the trip may seem to have been poorly planned from the start, nothing could be further from the truth. Mexican and Japanese officials had negotiated and researched everything—from the beginning of bilateral relations to the establishment of these settlers—for over a decade. In the process, Mexico became the first non-Asian country to sign an equal-terms treaty with Japan and the first in Latin America to establish a Japanese state-sponsored colony.

The contextualization of this case shows how Japanese and Mexican modernization projects and national imaginaries coincided in the interstices of Pacific imperial geopolitics at the end of the nineteenth century. Contrary to the Chinese case,

---

1. María Elena Ota Mishima, *Siete migraciones japonesas en México, 1890–1978* (Mexico City: El Colegio de México, 1982), 40–42; E. Mowbray Tate, *Transpacific Steam: The Story of Steam Navigation from the Pacific Coast of North America to the Far East and the Antipodes, 1867–1941* (New York: Cornwall Books, 1986), 30, 40–48.
2. Enrique Cortés, *Relaciones entre México y Japón durante el Porfiriato* (Mexico City: SRE, 1980), 78; Ota, *Siete migraciones*, 39–40; *Relación de la visita oficial a la zona de la colonia Enomoto de la Chiapas, sur de México* (Mexico City: 1958), 62. The latter source is a report written in Spanish and Japanese in 1958 by a group of Japanese researchers who visited Chiapas to interview the Japanese descendants still living there.
3. Cortés, *Relaciones*, 78–79; Ota, *Siete migraciones*, 42–43; Katsuhito Misawa Saito, "La colonia Enomoto de Chiapas. Estrategia expansionista y proyectos migratorios japoneses a fines del siglo XIX: el caso de México" (MA thesis, Universidad Nacional Autónoma de México/Facultad de Filosofía y Letras, 1982), 156.

discussed in the previous chapter, Japanese and Mexican diplomats recognized each other as allies in the face of common Western imperial challenges. Furthermore, both countries' racialized nationalist discourses allowed for the incorporation of Japanese immigrants not only as alien temporary workers, as was the case for Chinese, but also as more permanent colonists and industrialists. In addition, the recurrent contacts and trips by dozens of Japanese and Mexican diplomats, businessmen, and colonists aboard *Gaelic* and many other Pacific steamers highlight the fact that nations and empires have been created through movement, that is, that state bureaucracies and national populations do not remain static inside a delineated border but rather move around as a means to negotiate their interests and implement their personal and group projects.

Japanese Modernization and the Beginning of Relations with Mexico

During the second half of the nineteenth century, Japan underwent a massive transformation. It began as a relatively isolated society under the Tokugawa shogunate when it faced being subordinated to Western colonial powers and grew into an industrialized capitalist empire. The bulk of changes happened between 1868 and 1912, corresponding to Emperor Mutsuhito's rule, known as the Meiji era. Politically, the main reforms roughly consisted of the restoration of the figure of the emperor, who centralized power as head of state and ruled with the support of a cabinet led by a prime minister who oversaw a large bureaucracy; a privy council in charge of key policies;[4] a bicameral system mostly responsible for voting on tax and budget matters and whose house of representatives was elected by a limited pool of male citizens; and a powerful and modernized military responsible for keeping order at home and defending its geopolitical interests abroad, which eventually reproduced mechanisms of colonial domination in nearby countries. Economically, an agrarian reform centered on the privatization and taxation of lands financed a rapid industrialization under state control, where mechanical advancements were initially copied from imported machines and later developed more autonomously. Key strategic industries related to the military remained publicly run throughout the period, but the remaining enterprises, notably those in heavy industry, flourished in private hands, especially under the *zaibatsu*, roughly defined as financial groups that operated as family trusts or through tight alliances. Despite their independent appearance, the state continued to regulate, protect, and subsidize the latter. This inevitably led to a budget deficit that lasted until the First World War. However, beginning in the 1890s there was a sustained decrease in manufactured imports and an increase

---

4. A third important group, whose functions were not delineated in legal documents but were part of a nonwritten tradition, was the *genro*, formed by elder statesmen who became the closest and most powerful councilors to the Emperor. The Constitution of 1889 and the different European-influenced codes issued around it established the new regulations. Omar Martínez Legorreta, "De la modernización a la guerra," in *Japón: su tierra y su historia*, ed. Daniel Toledo et al. (Mexico City: El Colegio de México, 1991), 185–87.

in exports.[5] There were also major philosophical, educational, and cultural reforms that are beyond the scope of this book.[6]

The colonial expansion of Western empires in eastern Asia was the catalyst at the heart of the Meiji reforms. As mentioned in Chapter 1, in the summer of 1853, commodore Matthew C. Perry and the war steamers under his command landed in Uraga and demanded the opening of Japanese ports to trade and to replenish US vessels with coal and other supplies. A year later Perry's troops forced the shogun to accept the first of a series of asymmetrical international agreements signed between 1854 and 1874 with the United States, Peru, Hawai'i, and thirteen European nations—the so-called Treaty Powers.[7] They all contained eight features distinguishing them from the reciprocal conventions signed among these countries. First, they were unilateral, that is, the privileges granted to foreigners in Japan were not offered to Japanese abroad. Second, they forced the opening of two cities and five ports—the so-called Treaty Ports—for foreigners to trade and reside in.[8] Third, these expatriate communities could create their own sets of laws and regulations. Fourth, only diplomats were permitted to travel into Japan's interior. Fifth, the treaties had no expiration date. Sixth, customs duties were fixed at a low, 5 percent rate. Seventh, the most-favored-nation clause forced Japan to grant the same concessions to all Treaty Powers. Eighth, foreigners enjoyed extraterritorial privileges, that is, they were ruled by their own legislations and were exempt from Japanese justice.[9] The shogun's acceptance of these unfavorable conditions forced his demise. This situation precipitated the Meiji reforms, which responded to US and European expansionism by, among other aspects, learning about and adapting Western advancements to compete with them more equitably.

The Meiji reforms coincided in time and substance with the rise of Mexican liberalism under Benito Juárez and later General Porfirio Díaz Mori. Mexican elites identified with the Japanese desire for material progress as opposed to what they described as an alleged Chinese aversion to Western ideas and technological advancements. In the words of Francisco Díaz Covarrubias, head of the Mexican Astronomic Commission analyzed in Chapter 2, the Japanese "show a true eagerness to instruct themselves and a strong determination to introduce in their

---

5. Martínez, "Modernización," 173–215.
6. For a more in-depth overview of the political and cultural transformations in Meiji Japan, see Marius Jansen, *The Making of Modern Japan* (Cambridge, MA: Harvard University Press, 2000), 294–494. To learn how ordinary people and different localities experienced and participated in the transformations, see M. William Steele, *Alternative Narratives in Modern Japanese History* (London: Routledge, 2003).
7. The European nations were Austria-Hungary, Belgium, Denmark, France, Germany, Great Britain, Italy, The Netherlands, Portugal, Russia, Spain, Sweden-Norway, and Switzerland. Louis G. Perez, *Japan Comes of Age: Mutsu Munemitsu and the Revision of the Unequal Treaties* (London: Associated University Presses, 1999), 188.
8. They were Hakodate on the northern island of Hokkaido; Nagasaki in southern Kyushu; and Niigata, Edo (renamed Tokyo in 1869), Yokohama, Shimoda, and Hyogo (present-day Kobe) in central Honshu. J. E. Hoare, *Japan's Treaty Ports and Foreign Settlements: The Uninvited Guests, 1858–1899* (Kent, CT: Japan Library, 1994), vi, 6.
9. Perez, *Japan*, 47.

country all the big social reforms and material improvements that emanate from the Western nations' science and culture. Steam [technologies] and electricity have already planted their routes in the Japanese empire, while force will probably be necessary to achieve the same objective with Chinese, due to their aversion to all the elements of civilization."[10] Among the many coincidences between Porfirian and Meiji modernizations, two are of particular interest for this book: the promotion of economic growth using imported steam technologies and the desire to redefine relations with foreign powers on more equitable terms.[11]

In terms of steam technologies, both governments budgeted large sums for railroad construction and establishing steamship routes. Nevertheless, rather than commissioning foreigners for both tasks, as Díaz's bureaucrats did after failed attempts to finance a national steamship fleet as described in the previous chapter, the Japanese state controlled the undertaking, importing machinery and sponsoring local business conglomerates. The process involved copying prototypes, understanding the components, and finally building according to their needs. Consequently, Japanese quickly learned how to develop and operate their own steam technology. For instance, in 1860, only seven years after Perry's arrival, *Karin Maru*, a diminutive 292-ton vessel, became the first Japanese steamer to cross the Pacific. It did so in only thirty-seven days. *Karin Maru* had Dutch engines and its local commanders, Kimura Yoshitake and Katsu Rintaro, were initially advised by a group of US naval officers. On the return trip, the Japanese crew made it home safely without kowtowing to a foreign advisor.[12] This marked the beginning of a successful maritime tradition that would make Japan a leader in the construction of steamships as well as the first Asian country able to sponsor transpacific liners.[13] In contrast, the first steamer entirely built in Mexico dated from the early twentieth century and was assembled in Frontera, a Gulf of Mexico port, by the US-owned Tabasco, Chiapas Trading and Transportation Company. The small vessel was limited to river navigation. Some years earlier, a shipyard had been established in Veracruz and later moved to Campeche, but workers there could only repair, not fabricate, steamers.[14] In the Pacific, prominent businessman Joaquín Redo—who operated a small maritime company, among other enterprises—opened a yard in Guaymas in 1891 to repair Mexican war vessels, steam engines, and boilers.[15] There is no evidence that complete ocean steamers were ever fabricated in the country.

---

10. Francisco Díaz Covarrubias, *Viaje de la Comisión Astronómica Mexicana al Japón para observar el tránsito del planeta Venus por el disco del sol el 8 de diciembre de 1874* (Mexico City: Políglota, 1876), 126.
11. Roberta Lajous, *La política exterior del porfiriato* (Mexico City: El Colegio de México, 2010), 140.
12. Tate, *Transpacific Steam*, 22; Dana B. Young, "The Voyage of the Kanrin Maru to San Francisco, 1860," *California History* 61, no. 4 (Winter 1983): 264.
13. For more on this topic, see Chapter 6.
14. "Nuestra construcción naviera," *El Economista Mexicano* 34 (April–September 1902): 451.
15. Miguel Ángel Avilés, "A todo vapor: Mechanization in Porfirian Mexico. Steam Power and Machine Building, 1862–1906" (PhD diss., University of British Columbia, 2010), 146–47. See the entire work for a pioneering study of steam technologies in Mexico, particularly in Sinaloa.

In terms of foreign affairs, Mexican functionaries were the first to identify the potentiality of establishing relations with Japan to bypass European interests. For instance, in March 1881, Manuel Fernández Leal, then Subsecretary of Development after having participated in the 1874 Mexican Astronomical Commission, sent a letter to Ignacio Mariscal, Secretary of Foreign Affairs, which stated:

> now that our [silver] coin has considerably lowered its value in European markets . . . and before the governments of England, the United States and other empires manage to impose another coin, we could send our pesos [to Japan and China] in exchange for the varied products from those nations . . . which, due to their cheapness and our advantageous geographical location, . . . we could resell in other countries. Also . . . the [increasing] population density of those [Asian] empires could provide us, in the near future, with a secure market for our Pacific coastal products.[16]

Matías Romero, Mexican representative in the United States for most of the last decades of the nineteenth century and introduced in the previous chapter, was another key advocate. Since his arrival in Washington in 1882, he had sent letters to Mexico City requesting governmental permission to start a rapprochement with the Japanese.[17] After securing official backing, he contacted Takahira Kogoro, his Japanese counterpart, to express his country's interest in formalizing relations. Romero visited Takahira and his successor Jishii Terashima Munenori several times that year. He offered them literature and presents from Mexico to stimulate interest. Romero also kept in contact with potential US mediators, such as former President Ulysses S. Grant, who had visited and maintained a network of acquaintances in both countries.[18]

The Japanese initially responded with caution, fearing another unequal treaty, but by the end of the decade they chose Mexico as their springboard to negotiate foreign relations under equitable terms. This change in attitude happened in 1888, when a new foreign affairs cabinet, led by Okuma Shigenobu, came to power following a governmental crisis regarding treaty revision.[19] After years of failed attempts to renegotiate with the Treaty Powers, Japanese diplomacy under Okuma decided instead to concentrate on negotiations with Mexico in order to set a precedent of equal treatment with a so-called Western country, as, up to then, only China had signed such an agreement with Japan. On June 23, Mutsu Munemitsu, the new Japanese representative in Washington, met with Romero to start treaty negotiations and demanded that Mexico renounce the principle of extraterritorial jurisdiction. In his words, this would make Mexico "the first civilized nation to do justice to

---

16. Archivo Histórico Genaro Estrada-Secretaría de Relaciones Exteriores (AHGE-SRE), 44-6-47, 1–2. Translations into English of Spanish documents by the author.
17. AHGE-SRE, 44-6-47, 19–20.
18. AHGE-SRE, 44-6-47, 12–14, 33–38, 76, 85–96, 100–102.
19. For more on this crisis and the path followed for treaty revision, see Perez, *Japan*, chap. 3.

Japan."²⁰ Mutsu argued that Japan was ready to offer foreigners fair legal treatment after having renovated its justice system according to European principles. There was therefore no more need for the extraterritorial jurisdiction demanded of the country in the past.²¹

Mexico responded cautiously. Ignacio Mariscal, the Secretary of Foreign Affairs, wrote to Romero, "Mexico could not take the initiative in recognizing [Japan's right to refuse extraterritoriality] . . . because it is not yet a powerful trading nation and, by taking that initiative, it could offend the powers that today have treaties with Japan. For that matter, Mexico would reserve the right to take that step until another nation of greater mercantile importance took it first. . . . Meanwhile the treaty could be celebrated in more general terms."²² After the Japanese refused to eliminate the subject from the treaty, Romero was instructed to consult on the matter with Thomas Bayard, the US Secretary of State. Even though the treaty with Japan was important for the Porfirian economic and international project, the relationship with the United States, the country's first trading partner, remained a higher priority.²³

While they had been the first to force Japan to sign an unequal deal, US representatives now considered the treaty system to be unfavorable to their interests because the United States had lost economic primacy to other countries. General Alexander C. Jones, the US consul in Nagasaki in 1881, presented the argument as follows: "the United States opened the country to the commerce of the world. Yet England and France have reaped the fruits . . . while the United States, her nearest neighbor and best friend . . . virtually does nothing."²⁴ By 1883, of the 208 foreign firms established in Japan, 98 were British and only 39 belonged to US investors. Britain also surpassed every other country in shipping tonnage, with 724,355 tons compared with 374,617 tons of all others combined. In terms of cotton textiles, Japan imported over $5 million-worth from Britain, some $133,000 from France, and only $73,000 from the United States. The US figures for yarn and woollen textiles were even less flattering.²⁵ Therefore, to counteract British trading supremacy, many US diplomats favored revising the treaty system to open the totality of Japan's market—instead of the seven cities to which they had up to then been limited—in exchange for accepting Japanese jurisdiction. Yet previous attempts to do so had been blocked by other Treaty Powers' representatives.²⁶ Consequently, when Romero

---

20. SRE-AHGE, 7-18-18, part 1, 78–85.
21. SRE-AHGE, 7-18-18, part 1, 87–88.
22. SRE-AHGE, 7-18-18, part 1, 86–87.
23. SRE-AHGE, 7-18-18, part 1, 172–73. See Chapter 3 for an overview on Porfirian policies.
24. Cited in David M. Pletcher, *The Diplomacy of Involvement: American Economic Expansion across the Pacific, 1784–1900* (Columbia: University of Missouri Press, 2001), 167.
25. Pletcher, *Diplomacy*, 167–69. For an overview on the economic impact of Western firms in Asia for the period studied here, see Francis E. Hyde, *Far Eastern Trade, 1860–1914* (London: Adam and Charles Black, 1973).
26. Pletcher, *Diplomacy*, 166. See also Arthur P. Dudden, *The American Pacific: From the Old China Trade to the Present* (Oxford: Oxford University Press, 1992), 142–43.

requested Bayard's opinion on the subject on October 26, 1888, the latter responded that he "personally wished . . . to recognize the autonomy of Japan and to treat her under absolute equality; . . . [based on these principles, the United States] had celebrated an extradition treaty with the [Japanese] Empire, the first of its genre, and another one of friendship, trade, and navigation, yet the latter cannot be executed until the European nations recognize the same principles . . . in order not to leave the [US] citizens in an inferior condition to the subjects of the European nations."[27] On November 2, Romero finally obtained from G. L. Rives, the acting Secretary of State, the much-sought declaration that "the interests of the United States would not be affected at all if Mexico conceded Japan the reciprocity that she wants in regards to the [extraterritorial] jurisdiction."[28]

After the US Secretary of State's approval, negotiations accelerated in earnest. Mutsu and Romero signed the bilateral Treaty of Friendship, Trade, and Navigation in Washington on November 30, 1888. Both legislative powers ratified it within six months, marking the beginning of official relations. Of the document's eleven articles, the fourth and eighth were unprecedented. The latter recognized each nation's jurisdiction over visiting subjects. In exchange for this concession, article four stipulated that Mexicans could establish themselves and trade throughout Japan. However, they never had enough economic clout to realize this extraordinary privilege.[29]

In the following years, Mutsu, now as Minister of Foreign Affairs, led the charge to renegotiate Japan's unequal treaties. Great Britain, the strongest of the Treaty Powers, came first. Negotiations started in Whitehall's Foreign Office in September 1893. Aoki Shuzo, minister in Germany and England and Mutsu's close friend, represented Japan. At Mutsu's request, Hugh Fraser, the minister to Tokyo, would head the British team even though he was on leave in London. Most importantly, Fraser was Aoki's friend and was sympathetic to treaty revision.[30] Knowledgeable of Great Britain's hunger for increased markets and antagonism towards Russia, Japan offered to open up its interior and unite against Russian expansionism in Asia in exchange for reciprocal terms. In July 1894, after close to a year of discussions, they signed the revised pact; finally, a European power recognized reciprocal jurisdictional rights in an Asian nation. In the words of Aoki, "we were able to get a more or less convenient treaty. With this success, we can suppress the disgrace that lasted more than thirty years and were able to join the fellowship of nations, which is a great happiness for us."[31] The US rapidly followed and by January 1897 Japan had

---

27. SRE-AHGE, 7-18-18, part 1, 234–35.
28. SRE-AHGE, 7-18-18, part 1, 244–49.
29. Carlos Uscanga, "Hacia una contextualización histórica de las relaciones diplomáticas de México y Japón," *Revista Mexicana de Política Exterior* 86 (June 2009): 72.
30. Perez, *Japan*, 101, 107.
31. Misawa, "Colonia Enomoto," 105.

renegotiated all its unequal treaties.[32] Its authorities could now begin consolidating not only a national but also an imperial project across the Pacific.

Mexico and the Japanese Pacific Imperial Project

At the end of the nineteenth century, in the context of European and US imperialism in Asia, Japanese elites began forming an expansionist project in which emigration played a central role. Emigration was highly restricted prior to Perry's expedition, but once its ports were forced open, the country joined the international network of labor, capital, and transportation that facilitated the movement of peoples. Western foreign businessmen who, over the past few decades, had been profiting from Chinese contracted labor attempted to reproduce the scheme with Japanese workers. Starting in 1868, US merchants hired Japanese to work on overseas sugar plantations, particularly in Guam and Hawai'i. After complaints from numerous migrants, Japanese authorities placed restrictions on private contractors and fostered the colonization of northern Hokkaido. They also signed a convention to provide sugar plantations with laborers that saw over 29,000 farmers travel to Hawai'i on three-year contracts from 1885 until the agreement's end in 1894.[33] Around the same time, thousands of Japanese began migrating to Australia and the western provinces of Canada and the United States. In the case of the former, they were mostly pearl fishermen and farmers hired by English contractors to work in Queensland's sugar plantations.[34] In the case of the latter, they were mostly student-laborers, that is, young men with limited resources who traveled abroad to study (and escape conscription) but had to work to cover expenses. Between 1882 and 1890, they accounted for close to half of the passports issued by the Japanese government to travel to the United States.[35] The remaining half was composed of laborers, free emigrants, prostitutes, and students with governmental scholarships.[36] Further north in Canada, early Japanese immigrants mostly worked in sawmills, fishing, and canning industries in and around Vancouver, as well as Vancouver Island's Cumberland mines.[37]

For Japanese authorities, emigration not only served as a safety valve for landless peasants and as a path for youngsters to become educated and send remittances,

---

32. The whole process is explained in Perez, *Japan*, 137–88.
33. Eiichiro Azuma, "Historical Overview of Japanese Emigration, 1868–2000," in *Encyclopedia of Japanese Descendants in the Americas: An Illustrated History*, ed. Akemi Kikumura-Yano (Lanham, MD: Rowman & Littlefield, 2002), 32–33; Yuji Ichioka, *The Issei: The World of the First Generation Japanese Immigrants, 1885–1924* (New York: The Free Press, 1988), 3, 40–46.
34. Misawa, "Colonia Enomoto," 82–83.
35. Of the 3,475 passports issued, 1,519 were for private students. Ichioka, *Issei*, 8.
36. See Ichioka, *Issei*, chap. 2.
37. Audrey Kobayashi and Midge Ayukawa, "A Brief History of Japanese Canadians," in *Encyclopedia of Japanese Descendants in the Americas: An Illustrated History*, ed. Akemi Kikumura-Yano (Lanham, MD: Rowman & Littlefield, 2002), 151.

but also formed part of a larger, Pacific Rim-wide expansionist project. Some elites justified expansionism not only in terms of an alleged manifest destiny that the "civilized" nations of the world shared, but also as a way to defend Japan from the other imperial powers. According to this logic, the stronger the Japanese looked in the international arena and the larger the territory they controlled, the less vulnerable they would appear to the European and US militaries, who actively sought to colonize new Asian territories and dominate new markets.[38] In this context, Enomoto Takeaki, Minister of Foreign Affairs between 1891 and 1892, sponsored a series of explorations throughout the Pacific Rim to evaluate the possibility of establishing state-sponsored colonies abroad.[39] Contrary to the Hawaiian experiment, this project encouraged permanent settlers rather than temporary workers,[40] a situation that offered several advantages. First, it would allow them to create and control outposts throughout the Pacific that could eventually become geopolitically useful. Second, authorities would remain in control of the entire emigration process to prevent mistreatment by foreigners, and Japan would benefit from colonist-generated profits. Third, this process promised an outlet for the increasing number of discontented, jobless, and landless peasants who found themselves in the nation's cities.[41] Fourth, as xenophobic reactions emerged in western Canada and the United States, the government was eager to find new territories where its subjects were welcome.[42]

After quitting his post, Enomoto continued promoting emigration with the foundation of the Colonization Society in March 1893. Composed of members of Tokyo's economic and political elites, the Society's main objective was promoting colonization. It did so by generating and disseminating information and facilitating favorable conditions for migrants. From its inception, the Society considered Mexico as a potential place to initiate operations, as suggested in Enomoto's inaugural speech: "if you ask me which plan should be executed first, I would like to respond that . . . I have the idea to organize the first colony in [the territory of] our eastern neighbor, in the Pacific coast of the Mexican nation."[43] In this context Mexico gained a presence in the nation's public opinion as a possible destination for colonists. While the media praised Mexico as the first non-Asian nation to sign an equal treaty, some questioned the viability of establishing colonies there due to its

---

38. Eiichiro Azuma, *Between Two Empires: Race, History, and Transnationalism in Japanese America* (New York: Oxford University Press, 2005), 18–19.
39. He sent emissaries to the Philippines, the Malay Peninsula, New Hebrides, and Mexico. Cortés, *Relaciones*, 73; Misawa, "Colonia Enomoto," 120–23.
40. Many of the Japanese who went to work in Hawai'i never returned, yet the Japanese authorities originally conceived the program as destined only for temporary sojourners.
41. Azuma, "Historical Overview," 32. Azuma also mentions that with the development of modern medicine and public hygiene, there was an exponential growth of the Japanese population at the time.
42. The first public event against Japanese in the United States happened in the spring of 1900, yet the Japanese arrived on the west coast of English-speaking North America in the context of anti-Chinese movements, which often translated into hostility or mistrust against "Orientals" in general. Ichioka, *Issei*, 52–53.
43. Cited in Misawa, "Colonia Enomoto," 104.

proximity to a strong rival power, the US. For instance, in August 1890, the *Japan Weekly Mail* stated that "the United States will never tolerate a foreign country setting foot in Mexico" and proposed Oceania and Korea as more suitable areas for colonization.[44] Other newspapers favored the idea, as suggested by a November 1891 article in the Tokyo-based *Keizai Zasshi*. After describing Mexico's promising geography, climate, and soil, the paper advocated for Japanese immigration since the ally had demonstrated "friendship and justice to Japan when signing a treaty under the base of equality. Besides, the country had invited the Japanese to establish themselves within her territory."[45]

Mexico's media also celebrated the rapprochement. In the summer of 1889, most of the capital's newspapers praised the new relations. The prestigious *El Economista Mexicano* printed the entire treaty and presented a compilation of various US news articles that congratulated Mexico for its diplomatic achievement.[46] In the following years, it published Nipponophile articles on trade, maritime routes, and immigration.[47] Mazatlan's *El Correo de la Tarde* printed favorable opinions on potential Japanese colonists in the spring of 1892, writing, "their presence would teach many . . . how to live well with little and how to work to obtain a reasonable remuneration. . . . We celebrate [their arrival] because we find in it a secure and practical way to augment the population and, as a consequence, the country's productive forces, without any of the inconveniences offered by [other races]."[48] The media also lauded the opening of the first Japanese furniture stores in Mexico City in the late 1890s, La Japonesa, La Crisantema, and El Lirio Japonés, as well as the 1899 visit of five entrepreneurs to Mexico City and of several others interested in importing silk products.[49]

Racialized arguments on both sides of the Pacific supported the colonization project. As far back as the late 1850s, Manuel Siliceo, then Minister of Development, and Matías Romero, working in Washington, considered there to be a racial affinity between Mexican Indians and Japanese. They based their assumption on the alleged similarity of features—intense black hair and eye color; brown or yellow skin; short stature; and oblique eye shape—concluding that racial kinship would translate into an easy assimilation to Mexican society.[50] Likewise, a November 1891 article in

---

44. Cited in Cortés, *Relaciones*, 73.
45. Cited in Cortés, *Relaciones*, 74.
46. "México y el Japón," *El Economista Mexicano* 7 (February–July 1889): 244, 282.
47. See *El Economista Mexicano* on the following dates: vol. 11 (February–July 1891): 308; vol. 13 (February–July 1892): 86; vol. 14 (August 1892–January 1893): 45; vol. 15 (February–July 1893): 5, 147, 219; vol. 18 (August 1894–January 1895): 5.
48. Article reproduced in its integrity in *El Economista Mexicano* 13 (February–July 1892): 86. See also the article reproduced in vol. 15 (February–July 1893): 5.
49. Cortés, *Relaciones*, 68–69.
50. Moisés González Navarro, *Los extranjeros en México y los mexicanos en el extranjero, 1821–1970*, Vol. 2 (Mexico City: Colegio de México, Centro de Estudios Históricos, 1993–1994), 178. Matías Romero, who was a strong advocate of Chinese immigration as well, as seen in the last chapter, also saw a racial affinity between the Mexican Indian and the Chinese, but he was one of the few—if not the only one—that sustained this

Tokyo's *Keizai Zasshi* was supportive, basing its outlook on friendly diplomatic relations and the racial affinity between the peoples of the two countries, which would contribute to the immigrants' rapid acclimatization.[51] That same year the Japanese minister in Washington, after visiting Mexico City's Belén jail, found not only racial similarities but also a linguistic one: the term "huarache," meaning "rustic leather sandal," which was worn by the working classes in both countries.[52] This was but one of many allusions made in Japanese functionaries' reports.[53]

An additional argument favoring the immigration plan was the supposedly positive traits of the Japanese "race." This led many to support their presence not only as temporary contract workers—as was the case of Chinese, discussed in the previous chapter—but also as more permanent *colonos* and industrialists. As we saw in Chapter 2, the Mexican Astronomical Commission's leader, Francisco Díaz Covarrubias, pioneered the idea that "a Japanese colony would offer our people a healthy example of everything that can be achieved with perseverance, laboriousness, and economy, even in the most unfavorable conditions."[54] He also suggested that they could help introduce many industries that they had mastered, such as the production of silk, porcelain, lacquer, and cabinet making.[55] Díaz Covarrubias's ideas remained in the popular imagination years after his trip, as the *El Tiempo* newspaper reproduced large sections of his 1874 account throughout the summer of 1889.[56] Matías Romero concurred and suggested that, due to Japan's rapid modernization—a process he desired for Mexico—its inhabitants would be welcome not only as laborers but also as "colonists, farmers, and respectable people."[57]

The media's increasing support for the Japanese emigration plan coincided with the appearance of various foreign service reports following the formalization of relations in 1889.[58] Mauricio Wollheim, a Mexican diplomat of German origin, would become the most dynamic promoter of emigration and exchanges to Mexico.[59] He joined the diplomatic corps in 1883 then requested a leave of absence to travel around the world. After his 1889 arrival in Japan,[60] he became convinced

---

publicly. See, for instance, Matías Romero, "Inmigración china en México," *Revista Universal* (August 1875). Reproduced in Josefina MacGregor, *Matías Romero. Textos escogidos* (Mexico: CNCA, 1992), 474.
51. Cortés, *Relaciones*, 74.
52. *Huarache* is a Spanish word used in many parts of Mexico, but its origins come from Purépecha, a First Nation language still widely spoken in the central-western part of the country. Francisco Díaz Covarrubias also remarked on the similarity between Japanese workers' and Mexican Indians' footwear. See Díaz, *Viaje*, 123.
53. González Navarro, *Extranjeros*, 180, 183.
54. Díaz, *Viaje*, 129.
55. Díaz, *Viaje*, 129.
56. See, for instance, the following editions: July 9, 10, 11, and 19, 1899. See also Cortés, *Relaciones*, 72.
57. Cited in Francis David Peddie Robson, "La colonia japonesa de México y la Segunda Guerra Mundial" (MA thesis, Facultad de Filosofía y Letras/UNAM, 2005), 54.
58. Cortés, *Relaciones*, 74.
59. Cortés, *Relaciones*, 103.
60. In 1891, Wollheim was appointed to the legation in Japan as first secretary and over the following seven years he occupied the posts of business attaché, general consul, and minister. Cortés, *Relaciones*, 66.

that "it would ... be of mutual benefit, if a portion of the laborers who continually are engaged ... for the cultivation of lands in other countries was diverted to [Mexico]."[61] From his point of view, the two countries had complementary projects, as Japan possessed "an ever-increasing population, which ... renders emigration necessary. Mexico, on the contrary, could easily support ten times the actual number of her inhabitants," who could work in the lands that remain "still without cultivation, because of the scarcity of manual labor."[62] Wollheim wrote various commercial reports, imported samples of Mexican products into Japan, and put up an exhibit in 1890 with the purpose of boosting bilateral exchanges.[63] Yet commercial relations remained relatively insignificant at a time when no regular direct maritime routes existed.[64] Wollheim traveled back and forth across the Pacific on at least four occasions, researching and reporting on areas to settle and lobbying stakeholders.[65] Enomoto was one with whom he "agreed ... in every respect."[66] Therefore, when Enomoto decided to execute the plan for a colony in Mexico, Wollheim's dream had come true—and he joined forces not only as consultant but also as investor.[67]

The Japanese foreign corps was even more active in promoting trade and emigration to Mexico. In 1890, Foreign Minister Aoki Shuzo ordered Chinda Sutemi, the San Francisco consul, to travel to Mexico and report on the possibility of commercial exchanges. Chinda arrived in the capital in December and met with various officials, including Foreign Secretary Ignacio Mariscal, Secretary of Development Carlos Pacheco, and Mauricio Wollheim, prior to his appointment to Japan. Chinda sent his compilation of information and observations from his meetings to the

---

61. Archivo Personal Matías Romero (APMR), reel 69, f. 48781, 1.
62. APMR, reel 69, f. 48781, 1–2.
63. For instance, in Wollheim's 1896 report, he listed the following Mexican products as having a potential market in Japan: gold, silver, copper, lead, iron, cotton, sugar, honey, indigo, skins, hemp, jute, henequen, tobacco, beans, wheat flour, garbanzo, peas, wool, coffee, cigars, timber, and caustic soda. In exchange, Japan could export the following to Mexico: silk products, tea, rice, camphor, straw mats, porcelain, furniture, wool blankets, coal, bronze, ivory, paper, and certain timbers. "Informes Consulares. Tráfico al Japón," *El Economista Mexicano* 21 (February–July 1896): 278. See also "El tabaco mexicano en el Japón," *El Economista Mexicano* 21 (February–July 1896): 183; "El tabaco mexicano en el Japón" and "Las sedas japonesas en México," *El Economista Mexicano* 20 (August 1895–January 1896): 101, 305. Wollheim's diplomatic successor, Fidel Rodríguez Parra, continued with his legacy. Find an analysis of one of his reports in "Porvenir del comercio entre el Japón y México," *El Economista Mexicano* 33 (October 1901–March 1902): 325. See also Cortés, *Relaciones*, 66–69.
64. For instance, in the first semester of the 1896–1897 fiscal year, Mexico imported roughly $13,329-worth of merchandise from Japan and $23,023 from China—compared with the $13.1 million imported from the US, its main trading partner, and $3.6 million from England—and exported only $4,000 to China and nothing to Japan—in contrast to the over $36 million to the US and $7.6 million to England. This does not include the amount of silver coins exported to Asia. "Importaciones y exportaciones de México habidas en el primer semestre del año fiscal de 1896-7," *El Economista Mexicano* 23 (February–July 1897): 29.
65. Cortés, *Relaciones*, 66. The route followed was the usual one, from Yokohama to San Francisco on a PMSS vessel and from there either by train towards Sonora or by steamer to any of the Mexican Pacific ports, notably Salina Cruz. See, for instance, AHGE-SRE, L-E-1856, 159, 257.
66. APMR, reel 69, f. 48781, 1.
67. APMR, reel 69, f. 48781, 7; Cortés, *Relaciones*, 66.

Gaimusho, the Japanese Ministry of Foreign Affairs.[68] Tateno Gozo—another diplomat who arrived in Mexico via the US—followed, becoming the first accredited Japanese envoy in the country after presenting his credentials to President Díaz on June 16, 1891. Tateno commissioned D. W. Jones, the newly established Japanese legation's official translator, to research the country's state of affairs, particularly its Pacific coast. Jones had lived in Japan for six years and Mexico for a decade;[69] he would produce one of the most complete reports on the country up to that point in time. He underscored the Pacific's economic potential, particularly the zone between Mazatlan, Sinaloa, and San Benito, Chiapas, at the southern border with Guatemala.[70] The vast area possessed fertile soil to support export crops such as cotton, coffee, sugarcane, tobacco, and rubber. He rightly foresaw trade increasing with the inauguration of the Tehuantepec transcontinental railroad in three years' time. Furthermore, as coastal populations were scarce, Japanese migrants would be welcome. Jones suggested that Japanese and not foreign capital should organize the immigration plan. From his perspective, the colonists would not need to learn Spanish. In fact, by not mixing or competing with locals they would not have to confront xenophobia; Japanese culture would prevail. Success depended simply on choosing the right piece of land and bringing in healthy workers who could best cope with tropical diseases.[71]

Fujita Toshiro was another great source of information on Mexico. He was tasked with opening a permanent Mexico City consulate and with covering for Tateno's constant absences (the latter was simultaneously the country's representative in both Washington and Mexico City). Just like Chinda, Fujita came from San Francisco's consulate, arriving in Mexico City in October 1891 and remaining in his post for twenty-six months. In the spring of 1892, he and Jones guided Foreign Minister Enomoto's four-person commission—made up of Morio Shigejiro, Tsuneya Moriyuki, Takano Shozo, and Enomoto himself—to evaluate the Mexican Pacific coastline for a possible colonization project. They left Yokohama for San Francisco in early April.[72] Fujita and Jones met them in Benson, Arizona, and together they crossed the border at Nogales.[73] They then took the train to the port of Guaymas, Sonora, where they embarked on the small 600-ton steamer *Alejandro*, which by then was making semi-monthly voyages between Guaymas and Manzanillo, stopping at all ports.[74] They arrived in Mazatlan in early May. A local newspaper described

---

68. "El tratado con el Japón," *El Tiempo*, December 9, 1890, 2; Cortés, *Relaciones*, 64.
69. Cortés, *Relaciones*, 76.
70. To learn about the history of Chiapas, development projects, and their repercussions during the Porfiriato, see Sarah Washbrook, *Producing Modernity in Mexico: Labour, Race, and the State in Chiapas, 1876-1914* (London: The British Academy, 2012).
71. Cortés, *Relaciones*, 64, 74–76.
72. Cortés, *Relaciones*, 64–65, 76, 110.
73. Misawa, "Colonia Enomoto," 144.
74. *Commercial Relations of the United States with Foreign Countries during the Years 1885 and 1886*, Vol. 1 (Washington: Government Printing Office, 1887), 877. The Mexican government gave a $1,200 monthly subsidy to *Alejandro* for carrying mail from all ports between Guaymas and Manzanillo.

Fujita as "a pleasant and intelligent young man recommended to the Governor by the President of the Republic." Tsuneya was referred to as "a well-known writer in his country [... who] had recently published a very important treatise on emigration." The newspaper praised "everyone [... for their] gathered efforts to give prosperity to Sinaloa.... Even though for the moment [their investments] are only speculations, they will undoubtedly prosper tomorrow."[75] The group continued their trip south along the coastal states of Nayarit, Colima, Jalisco, Guerrero, and Oaxaca. Their itinerary involved meeting with local authorities, inspecting export crop haciendas, mines, and wastelands, as well as gathering information concerning foreign trade and farming.[76] Once back in Mexico City in early October, Fujita declared to the local press that he was "delighted with his trip" and believed that "Mexico would soon count on a large number of Japanese immigrants."[77] Before returning home, the group personally thanked President Díaz for his support, gifting him an ancient Japanese dagger. In return, Díaz gave them a tour of his private collection of arms. Jones and Fujita stayed in Mexico City and sent samples of Mexican products with the four returning members.[78] By early December 1892, Morio, Tsuneya, Takano, and Enomoto sailed to Japan. The Mexican newspapers congratulated them on their safe arrival and celebrated that "their report will be entirely favorable towards Mexico."[79]

Only six months later, the Gaimusho sponsored a new research expedition. Nemoto Tadashi, the Colonization Society's secretary, left Yokohama in July 1893 and arrived in Mexico City in August, where he interviewed various officials over a period of several weeks. In October he arrived by steamer at the port of San Benito, close to the Guatemalan border. For a month and a half, he traveled across the state of Chiapas, particularly within the Soconusco, a coffee export area. He then visited the lands and the railroad construction in the Tehuantepec isthmus. Back in Mexico City, he declared that he found "Chiapas [to be] the richest state" and that "the first Japanese colony in Mexico will probably be established there."[80] Manuel Fernández Leal, now Secretary of Development, urged him to buy land in Chiapas for the foundation of a Japanese colony before land prices went up.[81] In February 1894, Nemoto returned home to hand in his report. A few months later he was back at sea again, this time to explore the countries of Central America.[82]

In early 1894, two other Japanese arrived in Mexico City on official missions. Jisashi Shinamura left his diplomatic post in New York to replace Fujita as the

---

75. "La comisión japonesa en Sinaloa," *El Monitor Republicano*, May 27, 1892, 3:1.
76. "Comisión japonesa," *El Monitor Republicano*, August 31, 1892, 3:3; Misawa, "Colonia Enomoto," 144; María Elena Tovar González, "Extranjeros en el Soconusco," *Revista de Humanidades*, ITESM 8 (2000): 35–36.
77. "Misión japonesa," *El Monitor Republicano*, October 11, 1892, 3:5.
78. Cortés, *Relaciones*, 111.
79. "La comisión japonesa," *El Monitor Republicano*, December 29, 1892, 3:2.
80. "La colonización japonesa," *El Monitor Republicano*, December 4, 1893, 3:1. See also Ota, *Siete migraciones*, 36.
81. Misawa, "Colonia Enomoto," 149.
82. Ota, *Siete migraciones*, 37.

general consul. In contrast to Fujita, who traveled around the country and actively participated in the projects for establishing Japanese colonies, Jisashi mostly stayed in the capital and after ten months transferred out.[83] Watanabe Chiyosaburo arrived in March as a representative of Tokyo's Nippon Ginko's (Bank of Japan) Council of Administration. He only stayed for a few weeks to study the fluctuations of Mexican silver coins which still circulated in Japan.[84] Two years later the Nippon Ginko transferred from silver to the gold standard and Mexican silver imports to Asia started a steady decline.[85]

In August 1894, the Gaimusho sponsored yet another expedition. This time agricultural specialist Hashigushi Bunzo was sent to examine the properties that Nemoto had reported on in Chiapas.[86] Hashigushi had specialized in farming at the University of Boston and was the director of Tokyo's Agricultural University. He left Japan on August 22, 1894, taking the usual route from Yokohama to San Francisco and then aboard the Panama-bound milk run. He arrived in San Benito on October 5. Over the next forty days he visited various Soconusco area properties. He then took a steamer towards Salina Cruz, the recently inaugurated transcontinental railroad's terminus.[87] From there he rode the rails to its endpoint, the port of Coatzacoalcos. He continued to the port of Veracruz and connected with the train to Mexico City. On November 24, he met with Secretary Fernández Leal and a week later with President Díaz; by January of the following year, he had returned to Japan.[88] In his final report to the Gaimusho, Hashigushi concluded that, of all the lands he had examined, the 116-hectare plot in Escuintla, Chiapas, was the best suited for their colony. Its topography allowed for cultivating coffee in the highlands and grains in the lowlands. There was ample space for cattle to roam and its proximity to a river and the sea could support a fishing industry. Additionally, the terrain's proximity to the Tehuantepec railroad and other local lines, which were already under construction, made it well connected. He suggested that Japanese authorities act quickly in order not to miss a good opportunity—he had been informed that a British colonization company was also interested in the area and land prices were rapidly increasing.[89]

One more land inspection would take place. Kusakado Toraji and Nemoto arrived in the capital in August 1896 on their way to re-examine the parcels suggested

---

83. Cortés, *Relaciones*, 111.
84. "Un banquero japonés," *El Monitor Republicano*, March 7, 1894, 5:1; Cortés, *Relaciones*, 66.
85. Between January and October 1900, Mexico exported via San Francisco and Europe some $494,500 of silver coins to Japan and $23.2 million to China. Yet during the same period in 1901, these figures had drastically diminished to $96,000 and $12.1 million respectively, as part of the transfer from silver to the gold standard that was happening worldwide, and due to the political instability of China. "Remesas de plata al Extremo Oriente," *El Economista Mexicano* 33 (October 1901–March 1902): 182. See also "La nueva ley monetaria del Japón y la baja de la plata," *El Economista Mexicano* 22 (August 1896–January 1897): 193–94, 206–7.
86. He is also referred to as Fushitoshi Hashigushi in some sources.
87. For information on this transcontinental railroad, see Chapter 5.
88. Ota, *Siete migraciones*, 37–38.
89. Cortés, *Relaciones*, 77; Misawa, "Colonia Enomoto," 149–50; Ota, *Siete migraciones*, 38.

by Hashiguchi and to formalize their intention to buy land in Chiapas. Nemoto already knew the area as he had traveled around southern Mexico and Central America two years earlier. From Mexico City the pair took a train to Veracruz, where they transferred to a Coatzacoalcos-bound steamer. They then rode the Tehuantepec transcontinental railroad to Salina Cruz and sailed to San Benito. On arrival they hired local engineers to help inspect the Escuintla land parcel, corroborate measurements, and take soil samples, and finally they confirmed their desire to establish the first colony in Escuintla.[90] Once the business was done, Kusakado returned to Japan and Nemoto sailed for Peru.[91]

Just as Kusakado and Nemoto arrived in Chiapas, a peculiar Mexican troop on an around-the-world trip landed in Japan aboard *Zaragoza*, a steamer custom built in 1891 for the Mexican navy in Havre's La Forgés shipyard. It had begun its mission in April 1895 in Tampico, in the Gulf of Mexico. It sailed south and circumnavigated the Strait of Magellan, arriving on the country's Pacific coast four months later. After repairs in the Guaymas shipyards, it left for San Francisco in April 1896. The steamer touched Honolulu and finally arrived in Yokohama in early August, staying in Japanese waters for two months. A young Mexican sailor aboard *Zaragoza* held his time there in the highest esteem, describing his favorite site, the Yokosuka arsenal, as "admirable [since] the Japanese ordered . . . a few vessels from England or France with the purpose of building three or four of the same model."[92] The Mexican vessel then departed for Hong Kong, where it stayed for three weeks before continuing across the Indian and Atlantic oceans. *Zaragoza* returned to the Gulf of Mexico in July 1897, completing Mexico's first global circumnavigation.[93]

The time was ripe for both countries to broaden their horizons.

The Enomoto Colony in Chiapas

The negotiations to buy a 65,000-hectare piece of land in Escuintla started in the summer of 1896. Murota Yoshibuni,[94] the new Japanese minister in Mexico City who represented Enomoto, and Fernández Leal, the Mexican Secretary of Development, ratified the final version of the contract on January 29, 1897. It stipulated that

---

90. Misawa, "Colonia Enomoto," 153.
91. "Los enviados del Japón, *El Monitor Republicano*, August 21, 1896, 2:4.
92. "Notas de viaje. El buque-escuela Zaragoza," *El Imparcial*, May 4, 1897, 2.
93. Mario Lavalle Argudín, *La Armada en el México independiente* (Mexico City: Instituto Nacional de Estudios Históricos de la Revolución Mexicana/Secretaría de Marina, 1985), 115–18; Mario Lavalle Argudín, *Memorias de Marina. Buques de la Armada de México. Acontecimientos notables, 1821–1991*, Vol. 2 (Mexico City: Secretaría de Marina, 1992), 77–95. President Díaz underlined this accomplishment in his speech delivered during the opening session of the 18th Federal Congress, referring to it as a proof that the "Nation does not stop in her quest . . . [towards] progress." "Informe leído por el C. Presidente de la República al abrirse el primer periodo de sesiones del 18º Congreso de la Unión, Septiembre 16, 1896," *El Economista Mexicano* 22 (August 1896–January 1897): 93.
94. Murota remained the Japanese minister in Mexico between 1895 and 1900, when he requested to be removed from his post due to health problems. Cortés, *Relaciones*, 111–12.

Enomoto had bought a "national terrain ... in Escuintla, located in the Soconusco department in the state of Chiapas ... destined for colonization." For this purpose, Enomoto had to bring in at least one Japanese family of farmers per every 2,000 hectares. In exchange, he obtained a discounted price of 1.55 cents per hectare, payable in fifteen yearly installments. Additionally, all sorts of tax exemptions applied as long as Enomoto brought in the proposed number of colonists.[95]

Enomoto and the Colonization Society wanted to start immediately. For this purpose, they put Kusakado, who had just returned from Chiapas, in charge of the project. Since the logistics for gathering families was going to take time, Kusakado decided instead to rely on young men from his hometown, Mikawa, in the Aichi prefecture. He convinced twenty-one men from his hometown to emigrate and complemented the group with eight men from Hyogo, three from Miyagui, and two from Iwate. All were between nineteen and thirty-four years of age. Six of them, Kiyono Saburo, Muramatsu Ishimatsu, Ota Renji, Sugawara Kotoku, Takahashi Kumataro, and Terui Ryojiro, would travel as free emigrants, the rest as colonists. According to their contract, the Colonization Society would pay for their transportation and a monthly salary of $12. It would also provide them with medical insurance, lodging, and tools for working the land. In exchange, the men committed to work for ten hours a day from Monday to Saturday.[96]

The colonists left Yokohama aboard *Gaelic* on March 24, 1897. It was a brand-new British vessel of over 4,000 tons when it joined O&O's transpacific service in 1885. Back in 1874, O&O had become the second service to challenge the monopoly of the Pacific Mail Steamship Company's (PMSS) San Francisco–Yokohama route. The first had been the China Trans-Pacific Steamship Company (CTPC), created a year earlier by a group of British merchants living in Hong Kong with the help of imperial subsidies.[97] O&O was created by the owners of the Central Pacific and Union Pacific railroads, which controlled all rail services west of the Missouri River, after PMSS announced that it would no longer transport its Asian cargo east via its railroads because of their high fares. Instead, PMSS vessels would unload in Panama and transfer goods via the local—and cheaper—transcontinental railroad. The US railroad companies reacted promptly, joining forces and creating their own transpacific firm. O&O chartered three British steamers and started operations in 1874. However, just as with its British counterpart, O&O's challenge did not last. PMSS soon reversed its decision after coming to an agreement with the railroad companies. By the end of the century, PMSS had successfully co-opted O&O and both worked under the same management, offering weekly sailings between San Francisco and Yokohama. *Gaelic* became one of the most active ships of their combined transpacific fleet.[98]

---

95. "Contrato," *Diario Oficial*, February 18, 1897, 3.
96. Ota, *Siete migraciones*, 40–42.
97. See Chapter 2 for a brief description of CTPC. PMSS is amply addressed in Chapter 1.
98. Tate, *Transpacific Steam*, 30, 40–48.

The colonists arrived at the port of San Benito, Chiapas, on May 10 after having transferred in San Francisco to a PMSS vessel doing the milk run to Panama. Tono Kuraji and Enomoto Ryukichi awaited them. These were Society envoys who had been sent to make preparations for their arrival. Yet when they arrived, the immigrants felt nothing had been properly planned. To begin with, after close to fifty days at sea, they had to walk for twelve hours in the heat to arrive at Tapachula, the nearest city. After only two days of rest, they started their 100-kilometer walk towards Escuintla—a three-day hike through the jungle. They arrived at the onset of the rainy season, which meant they were too late to prepare the fields. Sparked by the first showers, the weeds had grown beyond control. The group's inexperience with tropical agriculture showed when no one was able to successfully plant coffee, the colony's supposed main export harvest. The men were also unsuccessful with the crops destined for personal consumption, which ended up being eaten by neighboring cattle and birds. Additionally, they were attacked by local ailments for which they were unprepared, such as malaria, heat exhaustion, and insect and snake bites. On top of that, Kusakado ran out of money and could not pay the men their promised salaries.[99]

The situation rapidly unraveled. In June, the six free emigrants separated to form their own company, Teiyu Gaisha. They bought 222 hectares from Kusakado and began planting corn, beans, and sugarcane as well as raising cattle. In the long run, these youngsters became quite successful because as free emigrants they had access to more capital and overseas support networks than the colonists.[100] Furthermore, most of them had university degrees, which meant that they had been exposed to the intricacies of traveling abroad.[101] At the time, it was not uncommon for Japanese academic settings to be imbued with ideas of striving and success abroad to increase personal and national wealth and prestige. Therefore, at the end of the nineteenth century, thousands of young, educated, middle-class Japanese left their country with a certain amount of knowledge about how to be successful abroad.[102] As for the colonists, ten fled Escuintla in July. Five of them arrived at the Japanese legation in Mexico City demanding that consul Murota arrange for their immediate repatriation. In early August, Kusakado traveled to Japan to inform

---

99. Cortés, *Relaciones*, 78; Chizuko Watanabe, "The Japanese Immigrant Community in Mexico: Its History and Present" (MA thesis, California State University, 1983), 17; Ota, *Siete migraciones*, 43.
100. Free emigrants differed from contract laborers or colonists in that the former did not sign employment agreements before leaving Japan. Instead, they usually received financial help to go overseas from innkeepers, emigration companies, or personal contacts. Ichioka, *Issei*, 54.
101. Peddie Robson, "La colonia japonesa, 58; Watanabe, "Japanese Immigrant," 19.
102. Azuma, *Between Two Empires*, 20–25; Hisashi Ueno, in his 1994 book *Mekishiko Enomoto Shokumin*, arrives at the following conclusions on the subject: the most successful of the Enomoto colony's members were those with higher educational levels, as they assumed leadership positions from the beginning. Even though they were idealists, they conducted their businesses from a realistic perspective. They were self-reliant and married locals, all of which helped them to succeed in Chiapas. Cited in Jesús K. Akachi, et al., "Japanese Mexican Bibliographic Essay," *Encyclopedia of Japanese Descendants in the Americas: An Illustrated History*, ed. Akemi Kikumura-Yano (Lanham, MD: Rowman & Littlefield, 2002), 222.

Enomoto of the situation. He requested his own dismissal and proposed to continue the venture only with free emigrants, as they were the ones prospering in Escuintla. After the meeting, Enomoto decided to put an end to the Escuintla disaster. He sent Kawamura Naoyoshi and Kobayashi Naotaro to Chiapas with the purpose of repatriating the colonists. They left Yokohama in December. Kusakado never returned to Mexico; he committed suicide after delivering his report to Enomoto.[103]

When Kobayashi and Kawamura arrived in Chiapas, the situation in the colony had improved. Some crops had started to prosper and not everyone wanted to leave. At the end of January, only nine immigrants returned to Japan;[104] the rest stayed in Escuintla. Supported by consul Murota, who also traveled to Chiapas to evaluate the situation, they decided to rebuild. In May 1898, Kobayashi wrote to Enomoto that some 404 hectares of land had been successfully planted with corn, cacao, and coffee.[105] Two months later, diplomat Mauricio Wollheim reassured investors of Escuintla's viability at the General Meeting of the Colonization Society in Tokyo. He announced that he was so confident in the new Kobayashi management that he "offered to defray the expenses at Escuintla, which are estimated at $645 per month, during the second half of this year." In exchange, he demanded the corresponding number of shares to be transferred to him once the half-year had passed and had proven successful. His proposition was accepted.[106] Two years later a report from the Mexican government stipulated that the Enomoto colony's "sugarcane crops had rendered positive results . . . they provide the colony with 100 tons of sugar per hectare. . . . The colonists have also started with the cultivation of rice, but they have not planted coffee, because they have not yet found the appropriate terrain. Corn and bean sowing have been minimal, but there are projects to expand them. At the same time, the number of cacao and rubber trees has increased."[107]

In 1901, Enomoto ceded his rights to the Escuintla terrain to businessman Fujino Tatsujiro. Some administrative changes followed. Fuse Tsunematsu, an agronomist from the University of Komaba, replaced Kobayashi as the colony's administrator. Misumi Sutezo became Fujino's representative to the Mexican government. Concerned with the colonists' health, Misumi hired his close friend Tsuneki Horita as resident doctor. He arrived together with Fuse in 1902. Doctor Horita returned to Japan in 1904, only to come back to Escuintla two years later. This time he brought his new wife and a close friend, Naraki, who became the local pharmacist.[108] Fujino himself decided to move to Chiapas, too, where he worked with a reduced number of colonists and was moderately successful at producing

---

103. Ota, *Siete migraciones*, 43–44; Watanabe, "Japanese Immigrant," 17.
104. Watanabe, "Japanese Immigrant," 18.
105. Ota, *Siete migraciones*, 45.
106. APMR, reel 69, f. 48781, 7.
107. Cited in Ota, *Siete migraciones*, 45.
108. *Relación de la visita*, 61.

rubber and coffee as well as raising cattle for a decade. In 1914, the Mexican government declared the original contract with the colony void.[109]

While the Enomoto project in Escuintla failed at creating a successful state-sponsored colony, it succeeded in triggering a prosperous and growing migration of Japanese to southern Mexico. For instance, Fuse, the colony's administrator for several years, bought a farm for himself and invited young Japanese to immigrate and work with him. He was successful at producing coffee, cacao, and beans as well as raising cattle. Fuse stayed in the area until 1932, when he traveled back to Japan and died.[110] The founders of Teiyu Gaisha went on to become successful businessmen who encouraged Japanese to migrate to Mexico and work for them. In 1901, they joined with other fellow countrymen and formed San-O, a business association that moved from farming to retail as its main source of income. Its initial $500 in capital served to start retail stores in the neighboring towns of Acacoyagua, Pueblo Nuevo, Escuintla, and Tajuco. In 1905, its capital increased to $11,348. New members joined and San-O was renamed the Mexican–Japanese Company, Cooperative Society–Nichiboku Kyodo Gaisha. Three years later, the Cooperative boasted eighty-three members, including twelve Mexican women married to Japanese men and twenty-one of their children. Among them was Yamamoto Asajiro, one of the original colonists aboard *Gaelic*, who was now married to María Santiago Cruz, a native of Chiapas. The couple had six children. In 1913, the Cooperative, represented by Terui Rioziro, bought the colony's original lands from Fujino Tatsuhiro for 300 pesos.[111] In 1916, the Cooperative's capital ascended to $200,000 and its list of businesses included two farms, Tajuco and Permuta; a vegetable field, a pharmacy, and an ice cream parlor in Tapachula; a mill, two stores, a pharmacy, and an electric company in Escuintla; two pharmacies in Huixtla and Tuxtla Chico; and a clock shop in Tonalá. The Cooperative's prosperity attracted a growing number of Japanese, who arrived in Chiapas to work for its various businesses. They came not only from Japan but also from the United States, Guatemala, and Peru. Their stay in Chiapas allowed them to earn money and start their own businesses in the surroundings or in the neighboring states of Oaxaca and Veracruz. The Cooperative also brought teachers from Japan and opened a school for the children of Japanese–Mexican couples. They were also responsible for the first Japanese–Spanish dictionary edited in Mexico. In 1923, the Cooperative dissolved, but its members remained in the area and continued working independently.[112] Among them is Javier Juárez

---

109. Ota, *Siete migraciones*, 45.
110. Cortés, *Relaciones*, 81. Many of the businesses described in this paragraph suffered damages during the Mexican Revolution. See Chapter 6 to understand the impact of this war.
111. Archivo personal de Javier Juárez Yamamoto, "Testimonio de la escritura de compra venta celebrada entre los señores Tatsuhiro Fujino y la Compañía Japonesa Mexicana, Sociedad Cooperativa, representada por Rioziro Terui," February 1913.
112. Ota, *Siete migraciones*, 46–49; Cortés, *Relaciones*, 80–81; *Relación*, 56–61.

Yamamoto, great-grandchild of Asajiro, who now runs the cattle and fruit farm Tajuko in the nearby municipality of Acacoyagua.[113]

Conclusions: Mexico and Japan in the 1890s

At the end of the nineteenth century, Porfirian and Meiji elites were both engaged in modernization projects that relied on steam technologies to create enough material wealth to enable them to achieve more favorable stature in the world of nations. Since they did not pose a geopolitical threat to each other, they worked as allies in the face of stronger imperial actors in the region. In effect, Meiji elites used Mexico to set a precedent of equal treatment that helped them mitigate the effects of the uneven treaties previously signed with the United States, Peru, Hawai'i, and the European powers. Additionally, Mexico actualized a space where Japan's fragile dream of Pacific expansionism through colonies and maritime routes could flourish. For Porfirian officials, relations with Japan represented an opportunity to make their aspirations of establishing prosperous links with Asia come true, after all previous attempts—discussed in the previous chapter—had failed due to the need to resort to the intermediation of Pacific imperial powers. This intermediation became less necessary once Porfirian diplomacy established official relations with an Asian nation for the first time in the country's independent history.

After signing the 1888 Treaty of Friendship, Trade, and Navigation, both governments promoted the movement of peoples across the Pacific to consolidate their national and, in the case of Japan, imperial projects. Both sent their diplomatic corps as a vanguard and, by the early 1890s, opened offices in each other's territories, so that they could mutually engage. Japan also dispatched numerous emissaries to explore the political and physical terrain for the establishment of colonies. The constant cross-ocean flux described in the chapter demonstrates that nations and empires are formed through movement beyond their geographical borders and, more specifically, that transpacific movement related to Mexico and Japan was important to the construction of Pacific geopolitics at the time.

Just as had happened with Chinese migrants, Japanese immigration to Mexico was also explained in terms of racialized arguments; the difference was that those arguments served to support the Japanese presence in more favorable terms than the Chinese. As seen in the previous chapter, Porfirian elites either rejected Chinese by assigning them negative qualities or, at best, found a place for them in their national imaginaries only as alien, cheap, hardworking laborers. In contrast, Japanese were sometimes given the role of industrialists, that is, leading forces in the country's economic development, because Porfirian circles identified with the rapid modernization process that Meiji bureaucrats crafted. Consequently, Porfirian officers found a place for Japanese immigrants not only as temporary contract workers, but

---

113. Conversation with Javier Juárez Yamamoto on November 10, 2023.

also as more permanent *colonos*. And to further stress the incorporation of Japanese in the Porfirian national imaginary, some even suggested a supposed racial affinity between Mexican Indians and Japanese populations. By linking Japanese to the ancestral roots of the Mexican nationality, Porfirian policy makers created a historical tie that legitimized the Japanese presence in Mexico, even though records show that they had far less contact with colonial Mexico than did the Chinese.

In addition, the evidence discussed in this chapter shows that once the colonization project in Mexico left the bureaucrats' desks and became a reality, it took on a life of its own. While the first colonists arrived in Chiapas aboard *Gaelic* as part of a bilateral plan to populate the area with hardworking farmers who could produce export crops, the project did not materialize as conceived. Japanese colonists ended up following their own paths without strict state regulation. Many prospered, especially those who moved from farming to retail, married locals, and formed families.

Finally, the case highlights that *Gaelic* and the many other steamships that circulated in the Pacific served as in-between places where people from different cultural backgrounds with various personal and political interests connected with each other. In contrast, the next chapter shows how varied interests could not only coincide but also collide jarringly aboard steamers.

5
Suisang, 1908

Double Vision—Chinese Migrants and the Body of the Nation

Map 5
Alfredo A. y Carlos Vega Schiafino Jiménez, *Carta general de vías y comunicaciones de los Estados Unidos Mexicanos* [General map of Mexican roads and communications]. Map. Mexico City: Secretaría de Comunicaciones, 1907. From Mapoteca "Manuel Orozco y Berra," República Mexicana 5. Accessed December 15, 2023. https://mapoteca.siap.gob.mx/cgf-rm-m26-v5-0285.

Suisang arrived in southern Mexico's Salina Cruz on the morning of May 14, 1908. It had left Hong Kong weeks earlier with some 20 first- and second-class passengers and over 500 in steerage. *Suisang* was one of a handful of chartered steamers from the China Commercial Steamship Company (CCSC), a Hongkongese venture created in 1903 to establish direct routes between Asia and Mexico. Its main business was transporting Cantonese in steerage. During a routine arrival inspection, the port's health delegate, Dr. Valenzuela, found close to 400 men, 10 from second-class and the rest from steerage, affected with trachoma, a chronic, contagious eye disease. He immediately sent a telegram to his superior. Dr. Liceaga, a member of Mexico City's Board of Health, received the communique and subsequently refused the infected steerage passengers permission to land; affected second-class travelers were detained temporarily before subsequently being released. CCSC's management refuted Dr. Valenzuela's diagnosis and protested at once.[1] While the captain's and the ship's surgeon's affidavits showed that during the last week of sailing some 110 passengers had shown symptoms of an eye disease, it was "simple conjunctivitis . . . which [had] disappeared after treatment."[2]

Over the next month, the dispute escalated into a diplomatic battle, once again involving—as in the case of the *Mount Lebanon*, discussed in Chapter 3—Mexican, Chinese, and British authorities. Meanwhile, some 400 Chinese men remained captive aboard the *Suisang* in the direst of circumstances. Miserable sanitary conditions combined with intense tropical heat to take a disastrous toll: mumps, beriberi, and tuberculosis began to spread. The original eye ailment—whether trachoma or "simple" conjunctivitis—was exacerbated.[3]

The analysis of this case reveals that health diagnoses involving Chinese in Mexico were not only medical but also ideological affairs. As the arrivals of Cantonese dramatically increased with the start of bilateral diplomatic relations and the advent of CCSC, so too did disputes over their place within the Mexican nation. The case also illustrates the evolution of the first steamship company that finally established direct regular routes between Asia and Mexico as well as the kinds of problems faced by transpacific passengers and the diverse outcomes depending on their traveling class and nationality. Finally, it exposes the importance of Salina Cruz in the Porfirian Pacific strategy as well as in the transpacific port network once it became the terminal of the country's first—and the continent's fourth—transcontinental railroad and, as such, developed into a locus for international disputes such as the one examined here.

---

1. Public Records Office–Hong Kong (PRO-HK), C.O.129/352, 342–43 and C.O.129/378, 102, 106.
2. PRO-HK, C.O.129/378, 105.
3. PRO-HK, C.O.129/352, 344, 354; C.O.129/378, 108.

The Establishment of Sino-Mexican Relations and the Beginnings of CCSC

As discussed in Chapter 3, Mexican diplomats enthusiastically approached their indifferent Chinese peers to initiate relations in the early 1880s. A decade later, the Chinese would change tack. Rather than seeing a bilateral treaty as a distraction from more pressing issues, they viewed it as a possible solution, particularly in regard to the deteriorating situation in North America. In 1892, while the westernmost province of Canada, British Columbia, continued imposing its $50 head tax on laborers' arrivals,[4] the US government extended its Exclusion Act for another decade.[5] Chinese diplomats thus opted for creating favorable conditions for Cantonese to settle in neighboring Mexico. Pressure mounted also from US-based Chinese businessmen, who sent a commission to meet President Porfirio Díaz in Mexico City that same year.[6]

Direct negotiations restarted in Washington in 1893 when the newly appointed Chinese representative, Yang Ju, contacted his Mexican counterpart, Matías Romero. A Chinese governmental commission then visited Mexico City to analyze immigration conditions.[7] In the ensuing years, talks stalled due to several factors, notably the discussions on extraterritoriality, Yang Ju's transfer to Moscow in 1896, and Matías Romero's sudden death in 1898. However, on December 14, 1899, the representatives in Washington, Manuel Aspíroz and Wu Ting Fang, jointly signed the Treaty of Friendship, Commerce, and Navigation between Mexico and China, with article 5 stipulating that Chinese could freely immigrate to the Mexican territory.[8]

Throughout the 1890s, several Chinese and Mexican businessmen participated in the lucrative enterprise of bringing Cantonese laborers to Mexico, but none with any ease. This was the case, for instance, of Salvador Malo, one of the owners of the failed Mexican transpacific steamship companies described in Chapter 3. He requested hundreds of workers, notably from the Hei Loy Company of Hong Kong, to participate in the construction of the Oaxaca and Veracruz state railroads and

---

4. The tax was increased to $100 in 1903 and then to $500, which was equivalent to two years of each laborer's wages. "Federal Head Tax," Government of British Columbia, accessed October 28, 2023, https://www2.gov.bc.ca/gov/content/governments/multiculturalism-anti-racism/chinese-legacy-bc/history/discrimination/federal-head-tax.
5. The Chinese Exclusion Act was extended indefinitely in 1902. It was repealed in 1943.
6. Vera Valdés Lakowsky, *Vinculaciones sino-mexicanas: albores y testimonios, 1874–1899* (Mexico City: UNAM, 1981), 107–8. See also Vera Valdés Lakowsky, "México y China: del Galeón de Manila al primer tratado de 1899," *Estudios de historia moderna y contemporánea de México* 9 (1983): 9–19; Jingsheng Dong, "Chinese Emigration to Mexico and the Sino-Mexico Relations before 1910," *Estudios Internacionales (Chile)* 38 (January–March 2006): 82; Mercedes de Vega et al., *Historia de las relaciones internacionales de México, 1821–2010*, Vol. 6 (Mexico City: SRE, 2011), 99.
7. De Vega et al., *Historia de las relaciones*, 71.
8. Valdés, *Vinculaciones*, 100–120. All the primary documents generated by Mexican diplomats during the entire process of negotiations for the treaty with China are found in AHGE-SRE, L-E-1983, L-E-1984, L-E-1985.

for whom he received commissions.⁹ After the treaty's ratification, several other merchants approached Wu Ting Fang for permission to bring laborers from Hong Kong to Mexico. They also pointed out that there was a large market for a direct maritime route between Hong Kong and Mexico because up to then "all the Chinese proceeding by steamer from Hong Kong . . . have to cross in the United States in order to arrive to Mexico. This being very inconvenient," they requested that "the Wai Wu Pu [the Chinese Department of Foreign Affairs] might be asked to notify His Excellency the British Minister to communicate to His Excellency the Governor of Hongkong that China and Mexico are now under Treaty relations and that therefore in the future there would be nothing to prevent Chinese travelling by Steamers running between the two countries."¹⁰ British confirmation came in October 1902, when the Under Secretary of State, Francis Bertie, concurred that "there appears to be no objection to permitting such emigration in the future."¹¹

The first group of businessmen to run a regular transpacific steamship operation between Mexico and Asia was comprised of twenty Chinese subjects living in Hong Kong.¹² They formed CCSC on November 1, 1902, with the purpose of inaugurating a monthly service between Hong Kong and Mexico. Initial capital amounted to nearly $1,000,000, allotted in 9,980 shares of $100 each. The board appointed Eng Hok Fong as the first president.¹³ He sailed to Mexico and, on January 28, 1903, signed an official contract to establish "a shipping steamship service between Hong Kong or other ports from China, Japan and the United States and the port of Manzanillo and, if the company's interests allow it, to Mazatlan as well . . . making at least eleven round trips per year."¹⁴ In exchange for free transportation of governmental communications, the company received various tax exemptions. Unlike the Compañía Mexicana de Navegación del Pacífico (CMNP) and the Compañía Marítima Asiática Mexicana (CMAM), CCSC was not to receive any subventions for bringing laborers, as passengers would be treated as free emigrants to avoid any objections from Hong Kong's British authorities.¹⁵ Yet there were steerage passengers who were contracted laborers upon arrival and the company's main revenue

---

9. Jason Oliver, *Chino: Anti-Chinese Racism in Mexico, 1880–1940* (Champaign: University of Illinois Press, 2017), 70.
10. PRO-HK, "Chinese Emigrants for Mexico," August 13, 1902, C.O.129/312, 186.
11. PRO-HK, "Chinese Emigration to Mexico," October 7, 1902, C.O.129/314, 410.
12. On the one hand, Chinese businessmen living in Hong Kong benefited from the support of a British government that needed their money and know-how to make the colony thrive. On the other hand, these capitalists maintained ties with their hometowns, enabling them to have unfettered access to mainland China's people and resources. John M. Carroll, *Edge of Empires: Chinese Elites and British Colonials in Hong Kong* (Cambridge, MA: Harvard University Press, 2005), 37–57.
13. "A new shipping line in Hong Kong," *The Hong Kong Daily Press*, December 4, 1902, 2.
14. "Contrato," *Diario Oficial*, February 17, 1903, 677. Translation of this and the upcoming documents in Spanish is by the author.
15. From 1853 the British government forbade the emigration of Chinese contract workers to countries outside the British Empire. José Jorge Gómez Izquierdo, *El movimiento antichino en México (1871–1934). Problemas del racismo y del nacionalismo durante la Revolución Mexicana* (Mexico City: INAH, 1992), 31.

stream depended on their import.¹⁶ The contract was to be effective immediately and had a five-year duration, extendable indefinitely at the end of each cycle until either of the signing parties objected.¹⁷

The Official Guide for Mexican Railways and Steamships, which offered a monthly schedule of all trains and steamers circulating in the country, listed the first of CCSC's operations in June 1903. It stipulated that the company would run monthly steamers from Manzanillo to San Francisco, Yokohama, and Hong Kong—with upcoming sailings on June 10, July 7, August 2, and August 25. The publication listed the following agents: Elliot and Lange in Manzanillo, J. V. C. Comfort in San Francisco, and J. S. Van Buren in Hong Kong.¹⁸ The monthly steamship itinerary bulletin from the Office of the Postmaster General complemented the above information in its August 1903 issue,¹⁹ affirming that the CCSC's steamer *Clavering* would dock in Manzanillo on August 2, arriving from Hong Kong, Moji, Kobe, and Yokohama. It would then leave on August 21 for San Francisco, Yokohama, Kobe, Moji, and Hong Kong.²⁰ The same itinerary was listed in subsequent months for the other three company steamers.²¹ As we can see, on their journey eastwards, CCSC vessels traveled directly from Asian to Mexican ports and, on the way back, they docked in San Francisco. This way the company offered its Chinese steerage passengers a direct service to Mexico without the hassle of a San Francisco stopover, which was the itinerary followed by the Pacific Mail Steamship Company (PMSS) and all other transpacific carriers,²² and where, due to the Exclusion Act, they would not be allowed to disembark. An additional advantage offered by the CCSC in its early days was its lower steerage price: $15 per person as opposed to the $50 charged by PMSS. Not surprisingly, during the first year of CCSC operations, Chinese immigration to Mexico doubled from around 1,900 in 1903 to 3,800 in 1904.²³ On the return trip, CCSC's steamers benefited from the large number of Asian passengers returning to their places of origin—whether voluntarily or after being expelled—with the San Francisco stopover.

---

16. PRO-HK, C.O.129/352, 412–13.
17. "Contrato," 677.
18. Van Buren had been working with the Pacific Mail Steamship Company (PMSS). Compañía de la Guía Oficial S.A., *Guía Oficial de los Ferrocarriles y Vapores Mexicanos* 3, no. 7 (June 1903). These guides are available for consultation in Hemeroteca Nacional's Fondo Reservado.
19. Dirección General de Correos, in Spanish.
20. Dirección General de Correos, *Movimiento probable de vapores*, August 1903, 8.
21. See the remaining 1903 and 1904 volumes of the Dirección General de Correos, *Movimiento probable de vapores*, available for consultation in the Biblioteca Miguel Lerdo de Tejada. CCSC operated with four large steamers. Grace Peña Delgado, *Making the Chinese Mexican: Global Migration, Localism, and Exclusion in the U.S.–Mexico Borderlands* (Stanford, CA: Stanford University Press, 2012), 94.
22. At that point, PMSS, Oriental & Occidental, and Toyo Kisen Kaisha were the companies that offered transpacific trips on a regular service between Asia and North America, and they all stopped at San Francisco. See advertisement in *The Hong Kong Telegraph*, January 13, 1904, 2.
23. Peña, *Making the Chinese Mexican*, 94.

Chinese and the Health of the Mexican Nation at the Turn of the Century

Some 60,000 Chinese arrived in Mexican ports at the end of the nineteenth and beginning of the twentieth centuries, particularly after the establishment of bilateral relations.[24] Many resettled without official papers in the United States. Some did so by crossing large stretches of unmonitored borderland or through frontier towns passing as Mexicans; others went by sea, stowing away in all sorts of vessels that circulated frequently between northern Mexico and California. Many crossed with the help of an organized system in which Californian Chinese associations as well as US and Mexican officers participated and profited.[25] The associations' agents would initially cover travel costs and logistics and later their clients had to pay back the fees, often at exorbitant interest rates. The associations distributed those newly arrived in Mexico to temporary jobs in both countries. Part of the client fee was distributed as "commissions" or bribes to Mexican and US officials, of which Ramón Corral, governor of Sonora between 1887–1891 and 1895–1899,[26] and vice-president of the country in 1904 and 1910,[27] was probably the highest-ranked recipient.[28]

Despite the allure of the US, there were thousands of Cantonese who decided to stay in Mexico. Official documents from 1895 and 1900—that is, prior to CCSC's establishment—listed the population of Chinese living in the country as 1,023 and 2,835 people respectively, but by 1910 the registered population increased to 13,203.[29] In the 1890s, most landed in Gulf ports such as Tampico, Veracruz, and Progreso rather than Pacific ports, but with the establishment of CCSC and of bilateral relations, this trend quickly reversed. For example, the 1895 registers show only seventy-seven Chinese immigrants landing in the Pacific ports of Guaymas, Mazatlan, Todos Santos, and Salina Cruz as opposed to 271 who entered on the opposite coast. In contrast, in 1901, of the 922 registered arrivals, 909 entered via Pacific ports and only 13 via the Gulf coast. By 1907, with CCSC fully functioning, there were 5,616 registered arrivals through the Pacific, the large majority via Salina Cruz, and only 247 by means of the Gulf ports.[30] During the first months of operation, most of the steerage passengers brought by CCSC were hired to work on the construction of the Central Mexican Railroad, particularly on the stretch linking the

---

24. Robert Chao Romero, *The Chinese in Mexico, 1882–1940* (Tucson: University of Arizona Press, 2010), 1.
25. The following books include multiple examples of this: Chao, *Chinese in Mexico*; Peña, *Making the Chinese Mexican*; Elliott Young, *Alien Nation: Chinese Migration in the Americas from the Coolie Era through World War II* (Chapel Hill: University of North Carolina Press, 2014); Erika Lee, *The Making of Asian America: A History* (New York: Simon & Schuster, 2015).
26. The state of Sonora borders Arizona.
27. Corral served in a number of other high posts, including governor of the Federal District and minister of interior. Ernesto de la Torre Villar, *Lecturas Históricas Mexicanas*, Vol. 2 (Mexico City: IIH-UNAM, 1998), 693, https://historicas.unam.mx/publicaciones/publicadigital/libros/lecturas/T2/LHMT2_067.pdf.
28. Chao, *Chinese in Mexico*, 1–3, 27–28, 43, 51.
29. Chao, *Chinese in Mexico*, 56.
30. Chao, *Chinese in Mexico*, 51.

port of Manzanillo—where they disembarked—with the city of Guadalajara.[31] The majority eventually moved north, and by the turn of the century there were already various communities of Chinese living in the states of Baja California, Sonora, Chihuahua, Coahuila, Nuevo León, Tamaulipas, and Sinaloa, but also in the south, in Oaxaca, Chiapas, and Yucatán. In Sonora alone, the state with the highest ratio of Chinese in the entire country, they quickly became the second largest foreign group, outnumbered only by US citizens.[32] While most started as contracted laborers working mostly in mines, agricultural haciendas, and railroads, many succeeded at establishing all sorts of small businesses, notably in retail.[33]

As seen in earlier chapters, Chinese had been the subject of negative stereotypes and prejudices since colonial times. The Mexican press in the northern states encouraged discrimination against them as the Chinese grew in numbers and flourished as independent businessmen. The press focused on three main arguments. The first suggested that they displaced the local workforce and therefore increased local unemployment or, at the very least, contributed to a decrease in salary levels. The second argued that they were inassimilable, too different from the Mexican ideal prototype, and in cases where they attempted to integrate by marrying local women, they would only degenerate the Mexican race with their allegedly inferior traits and mores. The third labeled them as polluting agents, as they often carried contagious diseases due to the unsanitary conditions in which they lived, and therefore that they represented a serious threat to public health and hygiene.[34] For the purpose of this case study, we will concentrate on the examples that illustrate the latter bias.

The first comes from Mazatlan, Sinaloa, where public opinion against the Chinese was among the most virulent in the nation, often pointing to Chinese residents as a source for epidemics in the city. As one of the main commercial hubs and maritime terminals for the prosperous northern states, Mazatlan served as an arrival port for many Chinese laborers, particularly those coming from nearby San Francisco. It was also the destination chosen by dozens of Chinese merchants who wanted to benefit from its booming economy. Mazatlan records show Chinese established there as early as 1841.[35] For decades they lived in relative harmony with the

---

31. Compañía de la Guía Oficial S.A., *Guía oficial de los ferrocarriles y vapores mexicanos* 3, no. 9 (August 1903): 10.
32. In 1890, Governor Ramón Corral reported that out of the 56,000 or so state inhabitants, 229 were Chinese residents, making them the second largest foreign group after the 337 US citizens. From that date until their expulsion in the 1930s, the Chinese in Sonora would rank as the first or second most populous foreign resident group. Evelyn Hu-DeHart, "The Chinese in Northern Mexico, 1875–1932," *The Journal of Arizona History* 21, no. 3 (Autumn 1980): 277.
33. The next chapter will discuss these types of experiences. See "Los chinos en México," *El Economista Mexicano* 7 (February–July 1889): 242; "Colonización china en Yucatán," *El Economista Mexicano* 12 (August 1891–January 1892): 208; "Pormenores sobre la inmigración china en Yucatán," *El Economista Mexicano* 12 (August 1891–January 1892): 224–26; "Irrupción china en México," *El Monitor Republicano*, December 24, 1891, 2; Gómez, *Movimiento antichino*, 58–64; Hu-DeHart, "Chinese in Northern," 275–312; Chao, *Chinese in Mexico*, 97; De Vega et al., *Historia de las relaciones*, 101.
34. Gómez, *Movimiento antichino*, 65.
35. Archivo Municipal de Mazatlán (AMM), Independiente, Presidencia, caja 10, expediente 2, July 29, 1841.

rest of the population. However, when the San Francisco-based Wing Wo Company began operations in 1886, it chose Mazatlan as its base for the import of Cantonese into Mexico. One of the first contingents arrived in April and was immediately labeled a plague by a Catholic newspaper.[36] By June there were some 220 Chinese lodged in two large houses on Mazatlan's main avenue, waiting to be transported to their intended final destinations in Salina Cruz and the interior of Sinaloa. A group of neighbors called the town council to complain about the unsanitary conditions they lived in and requesting their eviction to a remote location to prevent them posing a health hazard. Local doctors Zúñiga and Valadés were appointed to form a health commission. After visiting the premises, they filed a report full of Eurocentric scientific jargon supporting the argument that the Chinese posed a sanitary threat. For instance, they suggested that out of the 56.25 cubic meters of air that an individual required to live healthily according to recent European standards, the Chinese only had 18 per person. This, combined with several other unhealthy factors, made them conclude that: "1. The houses absolutely lack the necessary hygienic conditions ... 2. The immigrants' permanence in those houses is an imminent danger for the development of infectious and contagious diseases in the city. 3. This danger [would be] even more impending in the following rainy season."[37] The municipality then ordered the police to evict the Chinese, but the prefect responded that this was not possible because there was no law that specified the number of people allowed to live in one private home. In addition, he had received a letter from Gee Shoon, Wing Wo's representative, in which he promised to take any extraordinary measures required by the government. In order to force the eviction, the municipality responded by enacting a June 22 decree forbidding "the agglomeration of inhabitants in one single building [based on the opinion] of a health commission ... [as this] could compromise public health."[38] The case was concluded six months later when Ramón Ortiz,[39] one of the 150 Chinese remaining in the area, appeared at police headquarters accusing Wing Wo's agent of not fulfilling the workers' contract, which stipulated a payment of a peso per day worked. He claimed that they were instead left on their own with no work, no money, and no way to get by as they did not speak Spanish. They offered the municipality their labor in exchange for enough money to cover their basic needs while they waited for help from the Chinese authorities. The municipality accepted to pay their room and board and sent the bill to the Chinese consulate in San Francisco. Within three

---

36. "Los chinos en Mexico," *El Tiempo*, April 3, 1886, 3.
37. AMM, "Inmigrantes chinos. Se consulta si su aglomeración en las casas que ocupan son nocivas a la salubridad pública," Presidencia, caja 38, expediente 1, documento 10, May 1886, 8.
38. AMM, "Inmigrantes chinos. Se consulta si su aglomeración en las casas que ocupan son nocivas a la salubridad pública," Presidencia, caja 38, expediente 1, documento 10, June 24, 1886, s/f.
39. Many Chinese immigrants took Mexican names (either willingly or forced by their employers or the Mexican authorities, who could not pronounce their Chinese names) upon arrival in the country.

weeks the consul sent a $400 check and a written commitment from the Wing Wo Company to never again commit such a violation.⁴⁰

Guaymas, the main port of Sonora, similarly witnessed racist health disputes. For instance, in 1899, in the context of the signature of the treaty between Mexico and China, the local newspaper *El Tráfico* wrote a series of articles criticizing the presence of Chinese in the city and objecting to the arrival of more immigrants. The locus of contention was once again the laborers' barracks as well as Chinese-owned commercial establishments. After the newspaper targeted them as the source of the city's recent leprosy outbreak, the municipality appointed a commission to examine all Chinese establishments. Unlike in Mazatlan, the health authority found no evidence against them.⁴¹

With these antecedents, it is not surprising that upon initiating operations, CCSC would also be accused of bringing in aliens who were potential health hazards, especially given the laborers' ostensibly poor living conditions. In May 1903, a Mexico City newspaper described CCSC's first disembarkation of 500 Chinese passengers in Manzanillo, Colima, as follows: "The contractors threw them on shore and abandoned them. Our government has enacted very firm dispositions for the construction of barracks on the coast in order to lodge the immigrants; [for their] meticulous disinfection and [for] the establishment of quarantines. . . . The arrival of the Chinese caused great curiosity in Colima [city] and from there more than 1,000 people went to witness the disembarking of the Asians." The Chihuahua newspaper that reprinted this article made the following addendum: "we join those who demand that, were it not possible to stop the inrush of the people of the queue, let's at least have the greatest precautions and procure the greatest isolation of the yellow flesh."⁴² The dispositions mentioned by the newspaper article were apparently suggested by CCSC after seeing the anxiety produced by the disembarking of immigrants in such poor conditions.⁴³

Four months later, after an outbreak of bubonic plague in Mazatlan was blamed on Asian new arrivals,⁴⁴ the federal secretary of the interior temporarily banned their entrance. He later substituted this measure with the September 29 sanitary decree regulating their immigration, stipulating that:

---

40. AMM, "Inmigrantes chinos. Se consulta si su aglomeración en las casas que ocupan son nocivas a la salubridad pública," Presidencia, caja 38, expediente 1, documento 10, 1886, s/f.
41. *El Tráfico* newsclips from February to June 1899, reproduced in Humberto Monteón González and José Luis Trueba Lara, *Chinos y antichinos en México. Documentos para su estudio* (Guadalajara: Gobierno del Estado de Jalisco, 1988), 38–53.
42. "La inmigracion china," *El correo de Chihuahua*, May 31, 1903, 2, 4.
43. AHGE-SRE, 15-10-6, 1903–1904, "Inmigración japonesa. Medidas dictadas por la Secretaría de Gobernación," 34.
44. According to Pablo Yankelevich, "Revolución e inmigración en México (1908–1940)," *Anuario de la Revista Digital de la Escuela de Historia* 24, no. 3 (2011–2012): 42, https://doi.org/10.35305/aeh.v0i24.97, it was a Japanese vessel that was at the heart of the dispute.

1 . . . each immigrant or passenger,[45] before embarking [for Mexico], . . . [must have] a health certificate . . . legalized by the Mexican consul . . . [issued] no longer than two months prior to arrival. 2 . . . each vessel [must have] a disinfecting stove for clothes and luggage . . . as well as a Clayton machine to produce sulphuric acid to disinfect the ship and kill rats . . . 3 . . . in the port of disembarking, the vessel, the load and the luggage will be disinfected . . . passengers and immigrants will be subject to surveillance, isolation, quarantine. 4 Disembarking will happen in the port of Manzanillo [unless the number of immigrants is less than ten], where CCSC will build a pest house. 5 . . . additional caution measures by the Health Council should be observed. 6 The Health Council and its delegates will be able to detain the vessels for as long as necessary in order to carry through with the previous measures. These measures will be observed by the Chinese and Japanese immigrants coming from the ports of China and Japan.[46]

The Japanese government immediately complained, finding it discriminatory to explicitly target only the two Asian nationalities. After a few epistolary exchanges, the Mexican government eliminated all references to nationality. The decree expired after a few months, once the government considered there was no more "threat."[47] However, a new federal Sanitary Code, approved only months earlier, included hygienic regulations for those arriving by sea.[48]

Three weeks after the decree's publication, in October 1903, President Díaz created a federal commission, formed by Eduardo Liceaga, Rafael Rebollar, José María Romero, José Covarrubias, and Genaro Raigosa, to investigate the repercussions of bringing Chinese laborers to Mexico. After hearing various testimonies and gathering considerable data, the so-called Raigosa Commission concluded that their labor was needed despite them being considered inassimilable. Immigration thus should not rest in private hands but rather under the federal government's regulation based on the country's national interest. The following year, the commissioners met with Interior Minister Ramón Corral and gave a series of recommendations that included preference for Japanese over Chinese or Korean as well as a series of restrictions and regulations for the latter's temporary immigration. None were turned into laws and the commission was soon disbanded without a clear

---

45. "Immigrants" was the term used for the laborers destined to work in Mexico and for those traveling in steerage. First- and second-class passengers were usually referred to as "passengers."
46. AHGE-SRE, 15-10-6, 1903–1904, "Inmigración japonesa. Medidas dictadas por la Secretaría de Gobernación," 40.
47. AHGE-SRE, 15-10-6, 1903–1904, "Inmigración japonesa. Medidas dictadas por la Secretaría de Gobernación," 28–54.
48. The first Mexican Sanitary Code was issued in 1891, followed by a second version in 1894. The third was approved in December 1902 and came into force in 1903. José Quero Morales, "El derecho sanitario mexicano," *Revista de la Facultad de Derecho de la UNAM* (January–March 1963): 149–50; José Agustín Ronzón León, "Modernidad, sanidad y nacionalismo en el México porfirista. Una mirada historiográfica a través del código sanitario de 1894," *Tzintzun. Revista de Estudios Históricos* 75 (January–June 2022): 70; see also Peña, *Making the Chinese Mexican*, 102; Dong, "Chinese Emigration," 87–88.

explanation. It has been speculated that members of Díaz's inner circle who favored Chinese immigration, including Corral himself, may have had an influence.[49]

While the Raigosa Commission did not produce any laws, it triggered several publications. For instance, in Salina Cruz, Oaxaca—one of the CCSC's ports of entry—the state government performed a careful study. Local authorities reported some 200 Chinese living permanently in various municipalities, notably in Juchitán and Tehuantepec, close to the coast. Most were engaged in retail and several others worked as cooks, farmers, and in the laundry and hotel business. With the exception of Tehuantepec, where the authorities labeled them as lazy, dirty, and prone to stealing, all other districts with Chinese presence reported good or even impeccable behavior and not a single offense committed by them.[50] Commission members José María Romero and José Covarrubias published their own texts attributing a series of negative qualities to the Chinese, calling them xenophobic, fanatic, cruel, and uncivilized.[51] They argued that in the alleged quest for a strong, homogeneous Mexican race, Chinese were too different from the idealized Christian Western civilization and, as such, could not assimilate. For Romero, they could become a dangerous influence on the Indigenous peoples in Mexico as "due to the state of ignorance and misery in which they find themselves, the contact with the Chinese could produce a serious degeneration, as to avert their danger, the masses do not possess the moral strength [that characterize those with] economic wealth and intellectual culture."[52] Covarrubias, on his part, did not foresee any danger of degeneration because he thought that Chinese and Mexicans were so different that they would always remain separate. Both agreed on the economic benefit of immigration: in order to continue with its necessary industrializing progress, Mexico needed cheap labor that could adapt to harsh conditions, and the Chinese were considered to be perfect for that. For this reason, Covarrubias suggested that the government not systematically exclude them but "direct this immigration to the places that need it, reduce it to the most convenient terms and always keep in the hands of the government the direction of its movement."[53] This premise in a way guided the massive influx of Chinese to Oaxaca aboard CCSC vessels to complete a key project for the Porfirian regime: the Tehuantepec transcontinental railroad and its adjacent international ports of Coatzacoalcos and Salina Cruz.

---

49. Kenneth Cott, "Mexican Diplomacy and the Chinese Issue, 1876–1910," *The Hispanic American Historical Review* 67, no. 1 (February 1987): 82–84. Cott sustains that the commission resumed its work in 1906 but produced no formal reports.
50. Archivo General del Estado de Oaxaca (AGEO), Secretaría de Gobierno (Porfiriato), 120-12-73, 1903, Oaxaca, "Informes pedidos a los distritos acerca de los inmigrantes chinos."
51. Gómez, *Movimiento antichino*, 67.
52. Cited in Gómez, *Movimiento antichino*, 71.
53. Cited in Gómez, *Movimiento antichino*, 70. See also "La inmigración asiática," *El Economista Mexicano* 53 (October 1911–March 1912): 1–4.

The Tehuantepec Railroad

Building a canal or a railroad across the Tehuantepec isthmus—the narrowest stretch of the Mexican territory, measuring some 240 kilometers—had been an age-old, unachievable dream for local elites.[54] While the first survey dated from 1774, during colonial Spanish rule, the first concession was granted in 1842 by President Antonio López de Santa Anna to Mexican businessman José de Garay.[55] According to Santa Anna, the most "sure and effectual [means] for promoting national prosperity . . . [is by] making the Republic the center of commerce and navigation for all countries; . . . this must be the consequence of the establishment of an easy and short route from one ocean to the other. . . . By this enterprise in particular the nation will obtain revenues with which it cannot reckon at present, derivable from foreign trade, and immediately reap the advantages which must result from universal intercourse, when its soil shall become the emporium of commerce, and consequently teem with wealth and abundance."[56] Yet the unstable political climate made it impossible for the project to take off.[57] In fact, construction did not start until the 1880s despite Congress having granted over thirty different concessions to both Mexican and foreign investors over the decades.[58] The main reasons for the failures were geographical, notably the harsh topography; political, including civil wars, legislative instability, and foreign invasions; and economic, as the project

---

54. The Tehuantepec railroad has been the subject of multiple studies. For an introduction to the main historiographical debates around it, see Paul Garner, *British Lions and Mexican Eagles: Business, Politics, and Empire in the Career of Weetman Pearson in Mexico, 1889–1919* (Stanford, CA: Stanford University Press, 2011), 94–137.
55. While the decree called for a canal to be built, article 2 stipulated that in the section where it was impossible to build it, one could resort instead to a railroad and/or steam carriages. José de Garay, *An Account of the isthmus of Tehuantepec in the Republic of Mexico; with proposals for establishing a communication between the Atlantic and Pacific oceans, based upon the surveys and reports of a scientific commission, appointed by the projector, Don José de Garay* (London: J.D. Smith & Co., 1848), 28, 105. The geopolitical and economic importance of this transcontinental route dates from pre-Columbian times, as mentioned in Chapter 1. Nemesio J. Rodríguez, *Istmo de Tehuantepec: de lo regional a la globalización* (Oaxaca: Gobierno del Estado de Oaxaca, 2003), 2.
56. Garay, *Account of the isthmus*, 104–5.
57. Enrique Sodi Álvarez, *Istmo de Tehuantepec* (Mexico City: Talleres Gráficos de la Nación, 1967), 94.
58. The original concession owned by Garay was sold to the British firm Manning, Makintosh & Schneider in 1847. Makintosh was a lumber trafficker who served as the British consul in the area. Around the same time, as part of the war negotiations that led to the appropriation of half of the Mexican northern territory, US President James K. Polk offered to double the war compensation in exchange for exclusive rights of transit along Tehuantepec, a condition that the Mexicans did not accept. Nevertheless, the concession ended up in US hands when a group of New York bankers led by P. A. Hargous bought it from the British. In 1852 the Mexican government cancelled it, but five years later another US company, the Louisiana Tehuantepec Co., obtained it. In 1859, the Benito Juárez government signed the McLane–Ocampo treaty with the United States, granting them the exclusive and perpetual rights of transit in Tehuantepec, in exchange for $4 million, an amount that would finance the Mexican liberal army against the conservative forces in the midst of a civil war. The treaty did not come into effect as it was never ratified by the US Senate, in the context of a polarized political debate that would eventually lead that country to a civil war as well. Sodi, *Istmo*, 95–98; Rodríguez, *Istmo de Tehuantepec*, 2–3.

required large sums of capital that neither the government nor private investors were able to provide.[59]

As discussed in Chapter 3, rail expansion became an essential component of Porfirio Díaz's economic strategy, and within it, the construction of the Tehuantepec line occupied a central place. Díaz himself labeled it "a development of great importance and transcendence for the economic future of the country."[60] In addition to the economic benefits brought by the increase in trade, a Mexican transoceanic route would bring international prestige to Díaz as the country would join Panama, the United States, and Canada as the only nations with a transcontinental railroad.[61] For all these reasons, the Porfirian authorities channeled their energies into this task. In fact, the first loan raised in international financial markets went to subsidize the construction of the line, and the government invested 22.4 million pesos in subsidies to complete it, making it the costliest line in terms of public subsidy per kilometer of the entire railroad system,[62] and the most expensive engineering project undertaken in Porfirian Mexico.[63] The railroad finally opened in 1894 but with multiple defects, as evidenced by an internal report issued months after the inauguration: "the embankment is absolutely eroded, and none of it is adequately ballasted ... [and] the rolling stock is in truly dreadful condition."[64] British passenger Alfred Tischendorf described Tehuantepec as "one of the worst railways in the world. It was impossible to carry heavy loads on unballasted track or over wooden bridges that rotted under the stifling heat and torrential rains. The passengers who were willing to ignore the record of derailments and to brave the chances of contracting smallpox and yellow fever at the port cities, puttered along in swaying cars that occasionally reached a speed of thirteen to fifteen miles an hour."[65]

Aware of its shortcomings, the Porfirian government launched a bid to fix the deficiencies once and for all. British contractor Weetman Pearson beat all tenders, among which was C. P. Huntingdon, owner of the powerful PMSS. Pearson's bid was favored not only for economic reasons but also because of his close ties with Díaz's inner circle and his ability to manipulate the government's nationalist stance in his favor, claiming that he guaranteed "freedom from American control ... and the certainty that English trade and English ships would be predisposed to use a route controlled by Englishmen."[66] Members of the US government would resent

---

59. Garner, *British Lions*, 97.
60. Cited in Garner, *British Lions*, 101.
61. The transcontinental railroads in Panama, the United States, and Canada were inaugurated in 1855, 1869, and 1885 respectively. Panama was a province of Colombia when its railroad was first built but obtained its independence in 1903.
62. The average subsidy per railroad kilometer paid by the Díaz government was between 8,000 and 9,000 pesos. The Tehuantepec railroad received over 25,000 pesos per kilometer. Garner, *British Lions*, 102.
63. Garner, *British Lions*, 96.
64. Garner, *British Lions*, 104.
65. Cited in Romana Falcón et al., *Don Porfirio presidente nunca omnipotente: hallazgos, reflexiones y debates* (Mexico City: Universidad Iberoamericana, 1998), 110.
66. Garner, *British Lions*, 105.

the decision for years. In 1899, they asked for a copy of the Pearson contract as well as the nomination of one of their citizens as consulting engineer. Both requests were denied. In 1902 they requested an explanation as to why "the United States and its citizens should be so odiously excluded."[67] Just before the railroad's re-inauguration they even threatened to ban US merchandise from traveling along the Tehuantepec line.[68]

The initial contract between the Mexican government and Pearson and Son Ltd. was signed in 1898 and subsequently modified in 1899, 1900, and 1902. In accordance with the nationalist rhetoric that had surrounded the project since its earliest conception, it stipulated that it was a government venture and not a subsidy to a private enterprise. Yet the two parties were equal partners—each supplying half of the capital investment—and Pearson would be responsible for managing the Tehuantepec Railway Company (TRC) once it restarted operations. The partnership would last fifty-one years: during the first thirty-five, the Mexican government would receive 65 percent of the profits; after that it would progressively increase its share until it received the totality of the earnings. A special clause prevented Pearson from selling his shares to US businessmen.[69]

An important aspect of the partnership concerned the development of adequate port facilities at both ends of the line: Salina Cruz in the Pacific and Coatzacoalcos—renamed Puerto Mexico in 1907 to make it easier for foreigners to pronounce—in the Gulf of Mexico. After all, part of the failure of the first railroad had to do with the poor harbor conditions. One of the main obstacles to the project's progression was the lack of workers. Low population density combined with the reluctance of local inhabitants to work for the TRC made it necessary to bring in foreign workers. The majority came from Canton aboard CCSC vessels, which, by the end of 1904, had transported some 5,000 laborers, most of them to the Oaxacan coast.[70] Other workers came from Japan, Korea, Jamaica, and the Bahamas. Conditions were tough, with heavy rainfall, intense humidity and heat, and the proliferation of tropical diseases such as malaria and yellow fever. The construction of the port facilities in Salina Cruz posed more problems than those in Coatzacoalcos. Located in an open bay, the workers needed to erect a huge breakwater, build a one-kilometer-long dock, and create a small urban settlement there. Within ten years, what had been a small fishermen's village had turned into an international port with a stable population of some 6,000 inhabitants. The approximate cost of both harbors was 15,000,000 pesos.[71]

---

67. Garner, *British Lions*, 108.
68. Garner, *British Lions*, 108.
69. Sodi, *Istmo*, 128; Garner, *British Lions*, 107; María Paulo, *Origen de Salina Cruz* (Oaxaca: Familia Moreno Paulo, 1977), 67.
70. "Inmigración china," *El Correo de Chihuahua*, November 15, 1904, no page number. See also "La inmigración asiática," *El Economista Mexicano* 53 (October 1911–March 1912): 1–2.
71. Paulo, *Origen*, 62–71; Garner, *British Lions*, 107.

On January 20, 1907, General Porfirio Díaz himself officially inaugurated the port of Salina Cruz and the Tehuantepec railroad by operating a crane that transferred the sugar brought from Hawai'i by the steamer *Arizona* onto a railroad wagon. The arrival of the president and the party that ensued was described by the newspaper *El Tiempo Ilustrado* in the following terms: "the military rhythms of the bands, the cries and 'vivas' of the crowd, the whistle of the engines and the steamships in the port combined in deafening concert and brought joy and celebration to the hearts of all those present.... All the sacrifices that have been made seem to us today to have been amply rewarded with the satisfaction of having provided the world with a route across our territory which represents now and forever one of the major triumphs of the current president."[72] The celebrations lasted for a week and invitations were extended to the representatives of the United States, the United Kingdom, Germany, Spain, Russia, Belgium, Japan, El Salvador, Guatemala, and Cuba.[73]

Not Seeing Eye to Eye: *Suisang*, 1908

Once the Tehuantepec railroad and the Salina Cruz facilities opened for international carriers, many steamship companies moved their operations to this port, as it possessed the largest and most modern infrastructure between San Francisco and Panama as well as the most direct connection to the Atlantic. This was the case, for instance, of *Arizona*'s owner, the American-Hawaiian Steamship Company (AHSC), which specialized in transporting sugar from Honolulu to the east coast of the United States and with which Pearson had entered into relations to secure the large bulk of transcontinental operations it managed for TRC.[74] Another company was CCSC, which had been docking in Salina Cruz since 1904 and turned it into its main port of call by 1908.

By 1908 the CCSC had consolidated its transpacific presence. Its fleet of chartered steamers included *Marie, Daphne, Suisang, Glenesk*, and *Landkerscheiff*—most of them of British manufacture. With them, the company made a monthly voyage from Hong Kong to Salina Cruz with stopovers at Kobe and Yokohama, which took approximately five to six weeks to complete. The return trip went from Salina Cruz to San Francisco, Yokohama, Kobe, Moji, and finally Hong Kong, with occasional stopovers in other Mexican or Asian ports, depending on sufficient demand.[75] Their central business was in steerage. Between 1907 and 1908, each incoming CCSC steamer averaged 480 Asian passengers, roughly the standard for transpacific carriers of their size. Yet sometimes they hugely exceeded these figures, as did *Suisang*

---

72. Cited in Garner, *British Lions*, 116.
73. Garner, *British Lions*, 115; Paulo, *Origen*, 73–74.
74. Thomas C. Cochran and Ray Ginger, "The American-Hawaiian Steamship Company, 1899–1919," *The Business History Review* 28 (December 1954): 353.
75. Dirección General de Correos, *Movimiento probable de vapores*, February 1909, 12–13.

in early 1907, when it brought 737 passengers, and *Glenesk*, with 700 a few months later. Upon arrival, each traveler was subject to medical inspection, which often resulted in a few steerage passengers being denied entry as a result of suffering from some form of contagious disease. Before *Suisang*'s May 1908 docking, the largest number of CCSC passengers rejected by Salina Cruz's authorities had been forty-eight.[76]

Suisang left Hong Kong on April 2, 1908, and arrived in Salina Cruz on May 14, at 10:35 a.m., carrying 5 cabin, 24 second-class, and 518 steerage passengers on board. As part of the usual procedures, Dr. F. Valenzuela, the port's first health delegate, checked all passengers the next morning. Upon concluding his inspection, Dr. Valenzuela immediately sent a telegram to Dr. E. Liceaga, from the Superior Board of Health and one of the members of the 1903 commission reviewing the repercussions of bringing Chinese laborers to Mexico. Dr. Valenzuela informed him that he had found "341 [steerage passengers] affected with trachoma and seven [others remained] under observation . . . nine [second-class] suffering from trachoma and one from inguinal hernia, and five first class passengers in good health."[77] Dr. Liceaga ordered a new inspection to verify the initial diagnosis, which was performed two days later. This time Valenzuela found "three hundred and seventy cases of trachoma [amongst steerage] . . . and twenty more remain under observation. . . . Of the eleven second class passengers, ten are suffering from trachoma in a serious form." The latter were detained in the company's barracks "until they have decided to what place they are going in order that they may be duly kept under observation."[78] In the meantime all first-class and thirteen healthy second-class passengers were permitted to land. The 125 steerage passengers deemed in good health had been "lodged in the barracks, [had] been made to take baths, and their luggage [had] been disinfected; 121 pieces of luggage were fumigated and 132 treated with bi-chloride." On May 23 and 25, Valenzuela performed two more inspections in the *Suisang* and confirmed the existence of 371 cases of trachoma, "the majority . . . being serious ones. Amongst them were also two cases of mumps. [After] recommending to the ship's doctor the advisability of isolating the latter, he informed that there was no available place on board [for that]. The immigrants housed in the barracks were also isolated, and nineteen of those . . . were found to be suffering from trachoma. A further seven cases of mumps have appeared in the barracks. These are being isolated and the barracks in which they were housed are being disinfected."[79] After the usual ten-day quarantine for steerage passengers, Valenzuela proceeded to liberate 111 who were in good health and re-embarked 28, leaving on board the

---

76. PRO-HK, C.O.129/349, 312.
77. PRO-HK, C.O.129/352, 342–43.
78. PRO-HK, C.O.129/352, 343.
79. PRO-HK, C.O.129/352, 344–45.

Suisang "399 definitively rejected immigrants suffering from trachoma, and fourteen more who were rejected from the Steamship *Marie*."[80]

In terms of the sick second-class passengers, Dr. Liceaga ordered that officials "suggest to them that they enter the barracks, but do not force them . . . as we have no authority for taking this step, and advise us of the places to which they are going in order that they may be watched."[81] On May 26, ten of the eleven who remained in the barracks were set free, despite the fact that Valenzuela had diagnosed them with trachoma, four of them "in an advanced stage and very serious form."[82] Only one, suffering from mumps, remained isolated for a few more days and then was released on June 1. The only special measure taken with them was to record their names and destinations, listed as follows: Wohn Foo Wan, Durango, Railway Hotel; Wong Pun, Salina Cruz, house of Win Ong Wo; Ug Tye, Durango, house of Chang Lee; Leung Hoo, Mexico City; Jam Cherk Wang, Mexico City, house of Huong Fum Hai; Jam Took, Torreon, Railway Hotel; Chang Loing, Ciudad Juárez, house of Chong Kee; Woo Chang, Mexico City, house of Hog Huing; Lan Kee Char, Mexico City, house of Wach Chuc; Lam Ying Pang, Guaymas, house of Chin Cahn; and Kwong Wung, Salina Cruz, house of Chin Ong Woo.[83]

Meanwhile, CCSC's general manager, Mr. Leung Kam Ming, then stationed in Mexico City; the company's agent at Salina Cruz, Mr. Jesús M. Rabago; and the ship's surgeon, Dr. S. O. Netherton, protested to the Mexican Board of Health.[84] They claimed to have taken the necessary precautions required by law to prevent trachoma and other diseases from spreading on board. For instance, back in Hong Kong, right before departure, Dr. Grone, Health Officer of the Port, Dr. Paul, the company's resident doctor, and Dr. Netherton had examined all prospective passengers, and as

> evidence that such examinations were minute, careful and conscientious, out of 851 so presented for the Suisang 518 were accepted and 333 (about 40%) refused. . . . [Those accepted were sent to the Disinfecting Hulk,] fitted with the latest appliances for a thorough disinfection and to which all passengers and native crew . . . took a disinfecting bath and their clothes and personal effects were submitted to steam sterilization. After the bath they (still stripped) were again examined, their temperature taken and their physical condition noted especially in regard to signs of skin eruptions, enlarged glands, deformities, etc. [then] each passenger received a ticket marked "disinfected" and then proceeded direct to the ship in a launch and once on board they were not allowed to go ashore again. On the morning of the sailing all passengers were again passed, examined and counted by the Medical

---

80. PRO-HK, C.O.129/352, 345.
81. PRO-HK, C.O.129/352, 344.
82. PRO-HK, C.O.129/352, 345.
83. PRO-HK, C.O.129/352, 347.
84. PRO-HK, C.O.129/378, 106.

Officer of the Port, the Ship's Surgeon and the Boarding Officer in the presence of the Captain and the Co's Representative.[85]

Moreover, in compliance with article 49 of the Mexican Sanitary Code, right before departure each examined passenger received a bill of health signed by the medical officer of the port and the Mexican consul.[86] According to the company, with all these measures taken, it was impossible that a contagious disease such as trachoma had spread on board. The ship's surgeon, a thirty-nine-year-old graduate of the Jefferson Medical College in Philadelphia, claimed that "all the passengers were in good health and so remained till the last week of the voyage when 110 showed signs of simple conjunctivitis, and which disappeared after treatment some time after arrival."[87] After the Board supported Dr. Valenzuela's diagnosis, the company's representative made a new proposal to the Board and the Mexican minister of the interior: "In view of the great divergence in the results of the diagnoses of the Ship's Surgeon and of the Health Delegates of the Port, the Government be asked to consent to an expert oculist being sent to Salina Cruz . . . to be named by the Government but at the Company's expense, that all passengers be landed into the Co.'s barracks . . . [and] that the said expert's examination be conclusively accepted by both parties." Those deemed healthy or with "simple conjunctivitis" would be allowed to disembark while those diagnosed with trachoma would be "detained at the Co's barracks for treatment at the Co's expenses."[88]

The Mexican government did not accept the proposal, arguing that chronic diseases such as trachoma could take months to cure and it was therefore "practically out of the question to establish quarantine stations. . . . If, then, this is impossible, our only means of safeguarding the health of the Republic is to turn away all immigrants affected with this disease."[89] Additionally, Dr. Valenzuela was described as the country's expert on trachoma, having spent two years in the Far East studying the disease. There was therefore no need to bring in another doctor.[90]

CCSC's representatives therefore concentrated on disparaging Valenzuela's credentials and refuting his diagnosis. They deemed his first visit to the *Suisang* a "perfunctory examination, almost amounting to a farce," the results of which were suspicious as he kept increasing the number of diseased passengers every time he came aboard.[91] They accused him of being predisposed against CCSC and, in a

---

85. PRO-HK, C.O.129/378, 104–5.
86. Article 49 reads as follows: "Those persons who may desire to settle in the Republic or to be taken there in the character of immigrants, will only be allowed to enter the same, when they hold a certificate to prove their perfect state of health, issued by the Authorities of the place they come from and countersigned by the Mexican Consul. The certificates that may be issued two months previously to the date of arrival . . . will be of no value whatsoever." PRO-HK, C.O.129/378, 105.
87. PRO-HK, C.O.129/378, 105.
88. PRO-HK, C.O.129/378, 107.
89. PRO-HK, C.O.129/352, 370.
90. PRO-HK, C.O.129/352, 200.
91. PRO-HK, C.O.129/378, 106.

private exchange with Mr. Tower, the British consul in Mexico City, they suggested that the adverse attitude of the Mexican government was attributable to the influence of AHSC, which "desired to wrest from CCSC the coolie traffic."[92] In order to challenge Valenzuela's medical opinion, the company sent two Mexican doctors from Mexico City to Salina Cruz, Luis Buhot and Angel Vallarino, the former "being a graduate from the Institute of Ophthalmology and assistant to its Director Dr. Toussaint, and frequently called in for assistance by Dr. Chávez, one of the first oculists of this country, while the latter [was] professor of the School of Medicine in Mexico."[93] On June 9, after examining "412 individuals of Chinese nationality" aboard *Suisang*, Toussaint and Chávez reported "127 . . . with diverse ocular sufferings among which trachoma may be encountered, while of the [remaining] 285 we found to be free of all signs of trachoma."[94] A few days later, after making "microscopic and pathological examination[s]" of the tests made to the 127 aforementioned ill Chinese, they reported "that only about nine [were actually] suffering from trachoma."[95] Anticipating a similar problem happening with the next steamer about to land in Salina Cruz, the company ordered the two doctors to stay there.

Landratschieff, the next CCSC steamer to arrive, anchored in Salina Cruz on Friday, June 12, at 7 a.m. It had left Hong Kong on May 2 with 12 cabin and 550 steerage passengers on board.[96] Dr. Valenzuela came the next day to perform his examination. The company later complained that, once again, Dr. Valenzuela had purposely taken an additional day to perform his inspection, "in contravention of the Contract between the Government and the Co., which provided for an immediate boarding except on National Holidays or on account of stress of weather, and on these two occasions no such reason could be advanced." Valenzuela apologized and explained that this was due to problems he had with the tug boat.[97] To this, the company's representative, Mr. Camden, replied that he "could easily have visited the 'Landratschieff' in a small boat (as provided him by the Board of Health for such very purpose), his object being to delay the ship the next day being Saturday and he refused to work on Sunday."[98] Although Valenzuela performed his visit on Saturday, June 13, permission to disembark was not given until June 16. On that day, all cabin and 263 steerage passengers were allowed to land. The latter were detained in the barracks for the usual ten-day quarantine. After that, 122 were found free of diseases and, therefore, set free, while 136 were sent back to the vessel, having been found to be suffering from trachoma, making the total passengers rejected from the *Landratschieff* some 423.[99] Once again there was a discrepancy between Valenzuela's

---

92. PRO-HK, C.O.129/352, 252–53.
93. PRO-HK, C.O.129/352, 209.
94. PRO-HK, C.O.129/378, 109.
95. PRO-HK, C.O.129/352, 210.
96. PRO-HK, C.O.129/378, 102.
97. PRO-HK, C.O.129/378, 133.
98. PRO-HK, C.O.129/378, 110.
99. PRO-HK, C.O.129/378, 111.

diagnosis and the opinion of the ship's surgeon, Dr. Francis A. McOstrich, a graduate of the Royal University of Ireland. He argued that, in Hong Kong, all steerage passengers as well as the Chinese crew had undergone "three separate and distinct medical examinations by the port authorities on three separate days. In Salina Cruz [he contrasted] the examination occupied less than half a day. Just before the ship left, the passengers received their passports signed by the Mexican Consul and [the] Port Doctor, stating that the passengers were free from all traces of infectious disease[s] when leaving Hong Kong."[100] For him, there was, therefore, no possibility of trachoma spreading on his ship. To the contrary, he stated that "the health of [the] ship during voyage was excellent. . . . In my three years' experience of Indian and Chinese emigration I have never seen a more healthy crowd." The only problems he treated during the forty-one-day traverse, he claimed, were a dislocated shoulder, a carbuncle on a leg, one case of gonorrhea and bubo, one of cirrhosis, one of conjunctivitis, and one more of acute parotitis as well as several minor cuts and injuries.[101] Doctors Buhot and Vallarino, on their part, situated themselves somewhere in-between McOstrich's and Valenzuela's diagnoses, finding a "similar percentage of sick and healthy" as on the *Suisang*, that is, some 30 percent suffering from an eye disease and the rest healthy.[102]

The confrontation intensified after five Chinese escaped confinement. After CCSC requested a new examination, Dr. Liceaga ordered Valenzuela to return to the *Suisang*. He did so on June 15 and found five fewer passengers. The next day, after confirming that "the five missing persons [had] absconded," he ordered "the said vessel to anchor two miles off the Port until its departure has been decided upon, and . . . prohibited all communication with it to avoid further desertions." He also instructed that "no communication be permitted with the 'Landratschieff' until a proper watch be assured in respect to that vessel."[103] The company then appealed to the British authorities for help. The captain protested to the consul in Salina Cruz, "claiming that it was impossible to leave immediately as the ship needed water and provisions." The consul thus asked the Mexican authorities for an additional twenty-four hours, which were granted. Once the period expired, the captain continued to refuse to leave, claiming that the decision was "arbitrary and unreasonable."[104] Mr. Rabago, one of CCSC's representatives, requested the cancelation of the order, promising that the company would "establish a strict watch over the passengers . . . to prevent their escape, and in case any escapes should take place the Company [would] accept full responsibility."[105] He also requested the government to allow the "transfer of the Chinese immigrants from the 'Suisang' to the 'Landratschieff'

---

100. PRO-HK, C.O.129/378, 132.
101. PRO-HK, C.O.129/378, 132.
102. PRO-HK, C.O.129/352, 210.
103. PRO-HK, C.O.129/352, 348.
104. PRO-HK, C.O.129/378, 111.
105. PRO-HK, C.O.129/352, 350.

... because the former vessel having to call at Manzanillo to take on board 300 repatriated Chinese for Che Foo, [has not got] enough room to receive them."[106] Valenzuela told his superiors that he saw no point in maintaining a vessel "with 412 persons sick with a contagious disease on board, amongst whom symptoms of mutiny have been observed, and some of whom have already escaped."[107] In his view, it would be best if *Suisang* left Salina Cruz for good, without transferring any sick passengers to *Landratschieff*. And the same could be said for the latter. He also warned that if a mutiny occurred, a possibility suggested to him by the *Suisang*'s captain, "it would be a serious matter as the armed forces at our disposal in this Port would be insufficient to cope with such a rising,"[108] in addition to the fact that "the 412 immigrants on board would [make] their escape, and . . . eventually disseminate in all parts of the Republic the disease against which we are laboring to protect it from."[109] As a consequence the Board of Health replied that *Suisang* and *Landratschieff* should leave Salina Cruz. The message further explained that "the Nation is not concerned with them, nor consequently have the Sanitary Authorities anything to do with them. Treat them as if they were on high seas."[110]

CCSC again refused to accept the government's decision and left the vessels in Salina Cruz. On June 18, the Ministry of the Interior seconded Valenzuela and the Board of Health in obliging *Suisang* to lie two miles off the port; as a result, two days later the captain anchored the vessel in what he deemed "a most insecure and dangerous place, being virtually the open ocean."[111] As for the request to transfer the passengers to the *Landratschieff*, permission was granted provided that it was done outside the harbor limits to prevent further desertions that "would disseminate th[e] disease throughout the country, inasmuch as it was not possible to ascertain the destinations of the fugitives."[112] The company replied that this kind of transfer was impracticable and dangerous unless it was done inside the breakwater. Permission came after much pressing and, on July 8, 129 men were transferred from *Suisang* to *Landratschieff*. The former left for Manzanillo to pick up the passengers waiting there and came back to Salina Cruz on July 20 to try to force the disembarkation of the Chinese who had come from Asia in May.[113]

Throughout this time the poor sanitary conditions inside the vessels, combined with the intense tropical heat, had taken a disastrous toll on the health of the Chinese passengers. On June 30, after one of his renewed examinations on the

---

106. PRO-HK, C.O.129/352, 348–49. These men were contract workers from El Boleo, a French mining company established in Southern Baja California, which had arranged for CCSC to pick up its workers at Manzanillo and transport them to their homeland in China. PRO-HK, C.O.129/378, 111–12.
107. PRO-HK, C.O.129/352, 350.
108. PRO-HK, C.O.129/352, 349.
109. PRO-HK, C.O.129/352, 361–62.
110. PRO-HK, C.O.129/352, 349.
111. PRO-HK, C.O.129/378, 111.
112. PRO-HK, C.O.129/352, 361.
113. PRO-HK, C.O.129/378, 112.

Suisang, Valenzuela reported to have "found it in exceedingly bad sanitary condition, there being twenty-one persons suffering from beriberi and one from tuberculosis amongst the 409 trachomatous passengers on board. Three persons have died there within the last ten days from intestinal diseases and one from beriberi. If the vessel does not sail soon, [he warned,] I fear there will be a considerable mortality on board."¹¹⁴ He wasn't wrong. The cases of beriberi continued to increase, and nine more people ended up dying in Salina Cruz and over one hundred on the return trip to Hong Kong.¹¹⁵

The company sought a greater intervention from the British and Chinese authorities in Mexico. On July 17, the British minister of foreign affairs laid out a new proposal. It consisted of "appointing a Commission of medical experts, one to be nominated by the Government, one by the Company and the third by the two nominees, to examine and report on the passengers and to land those that were not suffering from trachoma."¹¹⁶ This time the Mexican government offered to form a commission with its own doctors, but this was rejected. CCSC's general manager then sought an interview with President Díaz himself. In it, Díaz affirmed that "under no circumstances would he change the Government's determination in refusing the men a landing, preferring rather to pay damages instead."¹¹⁷ The Mexican government claimed it was a priority to prevent that "the Chinese immigrants penetrate[d] into all parts of the Republic, [as] they would spread the contagion everywhere, so that the disease would at length become as prevalent in this country as it now is in China, in Japan and in Syria." Therefore, according to the experts from Mexico City's Academy of Medicine, the best possible course of action "suggested in these circumstances by hygienic requirements [leaned] rather to the rejection of persons unaffected with trachoma, than to the acceptance of those in whom the disease may be yet latent.... [As a consequence,] all persons believed to be suffering from trachoma, whether the disease be unmistakably developed, or its existence merely suspected, [were] to be refused admittance to the country."¹¹⁸

Convinced that there was nothing more to do, the company's owners decided to dispatch the steamers back to Hong Kong in August, after having spent close to two and a half months anchored in Salina Cruz.¹¹⁹ But it did so under protest. On October 22, the company sent a statement of claim to President Díaz for "expenses

---

114. PRO-HK, C.O.129/352, 354.
115. PRO-HK, C.O.129/378, 138.
116. PRO-HK, C.O.129/378, 113.
117. PRO-HK, C.O.129/378, 113.
118. PRO-HK, C.O.129/352, 369.
119. *Suisang* arrived in Salina Cruz on May 14 and left for Hong Kong on August 1. Between these dates, it left Salina Cruz for nine days, from July 11 to July 20, to pick up the 300 Chinese who were waiting in Manzanillo. *Landratschieff* arrived on June 12 and left on August 15. Taking into consideration that the trip from Hong Kong to Salina Cruz lasted approximately forty days, the rejected passengers from *Suisang* who made it back alive to Hong Kong had spent some 160 days inside a crammed vessel in the heat of the tropics, most of them suffering from an eye disease—whether trachoma or conjunctivitis—and later other contagious ailments. Those aboard *Landratschieff* spent two weeks less.

and damages incurred on account of the detention of its steamers ... through the rejection of its passengers" amounting to $3,076,184.87—which was later increased to $3,386,273.19. These included the extra expenses the company incurred for having *Suisang, Landratschieff*, and later *Marie* detained in Salina Cruz; the passages paid from Hong Kong to Salina Cruz and back by the "926 passengers that had been rejected and returned on board the three ships"; the profits that the company lost during the three months that the steamers did not circulate back and forth; the claims for the "130 passengers that died on board of the ships, compensation to be paid for their lives to their families, at $10,000 each"; all the passengers' loss of time and work; and finally an indemnity for rescission of contract.[120] The main arguments to support this claim were the following:

> A. That the 938 men refused landing by the Government did not have Trachoma ... B. That the examinations made, although repeatedly, by the Health Delegates, were in a perfunctory manner ... C. That the attitude of the First Delegate, Dr. Valenzuela, was one of extreme antagonism towards the Company ... D. That in not consenting to the proposals of the Co. re the dispatch of an independent expert oculist to Salina Cruz to examine the men, nor to that proposed by the British Minister for the appointment of a Commission of experts, the Government denied the Company justice and fair-play ... E. That the Delegate in deporting the men already landed into the Co's barracks ... exceeded his authority and usurped the authority of the Chief Executive, and such action virtually amounted to 'expulsion,' an action only within the province of the Executive ... F. That the individual bill of health ... was signed by Dr. Grone in his capacity as Delegate of the Mexican Government. This appointment was made by the Mexican Consul of Hong Kong on the departure of Dr. Valenzuela, who had been acting in that capacity ... G. That each and every one of the rejected passengers possessed the individual bill of health as prescribed for by Art. 49 of the Sanitary Code ... Excepting this, there is no other article, ... nor even in any of the laws of the country, that authorizes the refusal of a landing to any immigrant ... H. That the action of the Government was against their very Constitution promulgated 1857 [concerning the freedom of travel in the Republic and the right of the country to expel those foreigners considered 'pernicious'] ... Trachomatous persons cannot be considered as 'pernicious' ... I. ... that the action of the Government is contrary to the Treaty between that Government and Great Britain ... [and] the Treaty of 1899 between China and Mexico.[121]

The company included the following documents to support its claims: the contract between CCSC and the Mexican government, the sanitary provisions in force relative to immigrants from China and Japan, the instructions "which are to govern

---

120. PRO-HK, C.O.129/378, 113, 138. Even though the company claimed an indemnity for rescission of contract, some records indicate that it continued circulating, at least until the last year of the Porfirian administration. See AHGE-SRE, L-E-187, "Cuarta Conferencia Internacional Panamericana," 1910, 59. As for the Dirección General de Correos's *Movimiento probable de vapores*, it stopped listing CCSC's steamers as of 1910.
121. PRO-HK, C.O.129/378, 116–19.

the Delegate of the Supreme Board of Health in Hong Kong," and the affidavits of *Suisang*'s and *Landratschieff*'s surgeons.

The Mexican government responded to the accusations related to issues of health by extolling Dr. Valenzuela's credentials as an eye specialist and dismissing everyone else's. According to the Mexicans, Dr. Valenzuela,

> after having studied trachoma here in Mexico, with a specialist in the disease, went to Europe with the object of perfecting his studies of all diseases with which Chinese immigrants are likely to be affected, and subsequently continued the same in many places where Asiatic immigration is rife, such as Colombo, Singapore, Malacca, Hong Kong, Canton, Macao, Shanghai, Kobe, Nagasaki, and Yokohama. In these places collectively he had the opportunity of studying trachoma in about ten thousand different cases. These studies have enabled him to write a treatise on the subject, in which he has described with great exactitude the various stages of the disease.[122]

All this experience made him "exceptionally competent in the diagnosis of trachoma."[123] In contrast, neither of the two Mexican doctors hired by the CCSC, Dr. Buhot and Dr. Vallarino, could be "recognized as specialists in diseases of the Eye, and their competence as authorities on these diseases is neither attested by any Diploma nor recognized by the public at large."[124] Moreover, Dr. Buhot, presented by the company as a graduate of the Institute of Ophthalmology and assistant to its Director, could not have studied nor worked there "because there [was] no such institute in Mexico."[125] In the case of Dr. Grone, assistant to the medical officer of the port of Hong Kong and responsible for signing the individual bills of health approved by the Mexican consul, the Mexican government accused him of making "himself to be appointed Delegate of the (Mexican) Superior Board of Health in Hong Kong by our consul at that place, in order . . . to exploit the Chinese immigrants, from whom he was in the habit of collecting a fee of ten dollars for the issue of a Health Certificate."[126] Moreover, the Mexican consul had no authority to appoint medical delegates, which was an attribution only of the president himself. Therefore, the Ministry for Foreign Affairs would "doubtless hold this Consul responsible for having overstepped the limits of his powers." The Mexicans added that "CCSC must have been perfectly well aware [of this], through their Representative in Mexico . . . if he omitted to [know this], he and the Company are to blame for the Chinese immigrants in Hong Kong having accepted Dr. Grone as the accredited Delegate of the Board."[127]

---

122. PRO-HK, C.O.129/352, 364.
123. PRO-HK, C.O.129/352, 365.
124. PRO-HK, C.O.129/352, 366.
125. PRO-HK, C.O.129/352, 366.
126. PRO-HK, C.O.129/352, 365.
127. PRO-HK, C.O.129/352, 366.

This triggered a series of mutual dismissals and accusations that brought to light a possible network of corruption in the transportation of Chinese laborers to Mexico. It turned out that in October 1907, Dr. Valenzuela was appointed delegate of the Superior Board of Health in Hong Kong and therefore became responsible for approving the bill of health of workers wanting to immigrate to Mexico on CCSC's steamers. After his arrival, the examinations became more severe and the company faced an increasing percentage of rejections. CCSC found Valenzuela's objections exaggerated and, as proof, it offered a comparison between his percentage of rejections and that of Dr. Hough, who was in charge of examining Chinese passengers bound for the United States. While the latter had a 63 percent approval rate, the former only accepted 17 percent of all the men he examined. As a consequence, in the words of the British consul in Mexico, "Dr. Valenzuela and the CCSC in Hong Kong had differences of opinion, and these seem to have prejudiced the Doctor against the Company."[128] There seem to have been more than simple "differences of opinion," as Dr. Valenzuela had told Dr. Liceaga that "persons in Hong Kong, connected with the CCSC, attempted to suborn [him], and . . . failing this, an attempt was made to poison him,"[129] after which, in January, he decided to quit and leave for Mexico. Once returned, he was appointed the first health delegate at Salina Cruz. When his post in Hong Kong became vacant, Dr. Grone took it over in order to obtain illegal profits of up to $10 per immigrant he accepted, regardless of their actual good health, something that also indirectly benefited CCSC—or such were the conclusions made by the Mexican authorities. For them, an example that proved that his examinations were "not so scrupulously exhaustive as the complainants appear to imagine" was the fact that even the doctors hired by CCSC in Salina Cruz, Drs. Buhot and Vallarino, had admitted that there were some passengers with trachoma aboard *Suisang* and "they could not have contracted the disease except by contagion [of someone already sick aboard the vessel]."[130] To this Dr. Grone responded that he only charged the usual $2 fee per immigrant and the company added a letter from the Mexican consul in Hong Kong, F. D. Barretto, dating from November 1908, in which he stated that he had informed the Mexican government of Dr. Grone's appointment as a temporary replacement for Dr. Valenzuela in a dispatch dated January 20, without it being repudiated.[131] CCSC added that the Mexican authorities were inflating Dr. Valenzuela's credentials as the list of places where he supposedly studied trachoma were "simply the places at which [he] called on his way out."[132]

As for CCSC's accusations that the Mexican government had violated its own laws and international treaties by scrutinizing and expelling the Chinese immigrants

---

128. PRO-HK, C.O.129/352, 251.
129. PRO-HK, C.O.129/352, 201–2.
130. PRO-HK, C.O.129/352, 367.
131. PRO-HK, C.O.129/349, 314–15.
132. PRO-HK, C.O.129/349, 304.

arriving at Salina Cruz even though they possessed a health certificate signed at the port of departure, the government responded: "This assertion is unsustainable, for it is an axiom of [the] International Sanitary Police in all countries, that Health Officers have the right to examine not only immigrants, but also all passengers and crews ... when there is reason to suspect that they are carriers of infectious diseases ... after so long a voyage ... persons who started in good health might easily have developed some disease enroute."[133] Moreover, "no expulsion of Chinese subjects from Mexican territory took place, but merely a re-embarkation of immigrants who had been placed under observation in the barracks, after confirmation of the existence of suspected disease."[134] It was extremely important to prohibit the entry of people suffering from trachoma because the disease was "especially liable to spread rapidly amongst the poorer classes, who live huddled together and disregard the most elementary principles of hygiene. Seeing that a large proportion of our population consists of such people it was necessary to adopt special measures to prevent a hitherto almost unknown disease from becoming endemic in Mexico."[135] Additionally, the Mexican government could not be held responsible for the loss of time and lives of various Chinese passengers because "Suisang should have left the Port as soon as the separation of the healthy passengers from those who were to be rejected on account of disease, had taken place." If this vessel as well as the other two remained in Salina Cruz for so long, it was "not for the purposes of the Sanitary Authorities, but because it was convenient for the interests of CCSC," which kept requesting renewed health inspections.[136]

In December 1908, in what seems to have been a direct consequence of the *Suisang* controversy, the Mexican government enacted its first-ever Immigration Law. On the one hand, it included the liberal principle of "equality of all countries and all races, not establishing a single special precept for citizens of a particular nation, nor for individuals of a specific race."[137] On the other hand, it forbade entry to those carrying diseases that were more prevalent in Asian ports or that had been historically attributed to Asians in Mexico, including trachoma. The restrictions were enumerated in article 3 and included not only specific ailments but also "any chronic communicable disease." In addition, it forbade entry to those suffering "mental derangement," epilepsy, old age, or "any mental or physical defect that made people unfit for work or become a burden for society" as well as to anarchists, beggars, prostitutes, pimps, teenagers under sixteen traveling without a guardian,

---

133. PRO-HK, C.O.129/352, 367.
134. PRO-HK, C.O.129/352, 364.
135. PRO-HK, C.O.129/352, 368.
136. PRO-HK, C.O.129/352, 359–60.
137. Pablo Yankelevich and Paola Chenillo Alazraki, "El Archivo Histórico del Instituto Nacional de Migración," *Desacatos* 26 (January–April 2008): 30, https://desacatos.ciesas.edu.mx/index.php/Desacatos/article/view/535/401.

and offenders who had committed crimes punishable by more than two years in jail by Mexican law.¹³⁸

CCSC's claim, on its part, followed a long bureaucratic process that met no end during the remaining years of the Díaz administration. Under the advice of the British consul in Mexico, it was interposed as a private claim on October 22, 1908. On April 27, 1909, CCSC presented a corrected statement and added a series of vouchers and proofs addressed to the Ministry of Finance. On December 27, and again on April 25, 1910, the company wrote to the Mexican government demanding a settlement but obtained no official response. Rather, the authorities privately informed them that "the Government would resolve on the matter as soon as the election was over and the President re-inaugurated."¹³⁹ What neither Díaz nor CCSC were expecting was a set of popular uprisings that ended up undermining the Porfirian administration and forcing Díaz to resign and leave for exile. In May 1911, just as Díaz was preparing to depart for Europe aboard the steamer *Ypiranga*, the company's representative in Mexico sent the following telegram to CCSC headquarters in Hong Kong: "Claim cannot be settled in a friendly manner, the reason is revolution. Have left the matter in the hands of Minister (British). You must do it immediately, make application to the Governor of Hong Kong to telegraph London to telegraph instructions direct to Minister here for diplomatic action."¹⁴⁰ The governor of Hong Kong, Sir Frederick Lugard, agreed with the petition and cabled the Foreign Office (F.O.), which inferred that the revolution offered "an opportunity of reopening and pressing a claim which was previously incapable of settlement."¹⁴¹ After examining the case, the F.O. considered the CCSC's claim to be "no doubt exaggerated" and deemed the company "foolish (or clever) enough to allow the [rejected passengers] to disperse at Hong Kong without being again medically examined although it was stated that they would be so examined,"¹⁴² yet it gave its support by suggesting to the Mexican government "the advisability of appointing a representative to discuss . . . the amount of compensation which it would be reasonable for the Government to pay."¹⁴³ This idea was rejected in September 1912. On January 9, 1913, Francis Stronge, the British representative in Mexico, sent a note to the Mexican subsecretary of foreign affairs stating that "His Majesty's government . . . instructed me to urge your Excellency to submit the dispute with the CCSC to arbitration."¹⁴⁴ There is no further documentation to suggest that this procedure ever took place.

---

138. Yankelevich and Chenillo, "Archivo," 31; Mónica Palma Mora, "De la simpatía a la antipatía. La actitud oficial ante la inmigración, 1908–1990," *Historias* 56 (2003): 64–65, https://www.revistas.inah.gob.mx/index.php/historias/article/view/12952.
139. PRO-HK, C.O.129/378, 115.
140. PRO-HK, C.O.129/378, 115.
141. PRO-HK, C.O.129/377, 160.
142. PRO-HK, C.O.129/377, 161–62, 254.
143. PRO-HK, C.O.129/377, 253–54.
144. PRO-HK, C.O.129/377, 146–47.

Conclusions: Mexico and China in the 1900s

The signature of the Treaty of Friendship, Commerce, and Navigation between Mexico and China in 1899 allowed their citizens to circumvent the Anglo monopoly of two of the most lucrative transpacific businesses of the time: that of transoceanic steamship runs, dominated up to then by PMSS, and the transportation of Cantonese to the Americas, controlled by British authorities. While there had been previous attempts, described in earlier chapters, their successes had been limited, isolated, and irregular. CCSC was the first company to succeed in bypassing San Francisco and running regular direct trips between Asian and Mexican ports, and, as such, it had a stronger and more enduring impact on transpacific travelers to and from Mexico.

The establishment of bilateral relations and the existence of CCSC were made possible by a coincidence of multiple interests, actors, and circumstances. For Chinese authorities, Mexico became a much-needed escape valve for their southern populations seeking to emigrate in search of better economic opportunities than those provided at home, at a time when tighter immigration restrictions were being imposed by the US and Canadian governments. For Hongkongese businessmen, CCSC allowed them to profit from the transportation of Cantonese in steerage, the source of the company's central income. For San Francisco-based Chinese associations, the direct route to Mexico served to increase profits with the high interest rates and commissions charged to the Cantonese who immigrated under their patronage and surveillance. For Cantonese travelers, the new direct destination provided job opportunities as well as the possibility to join families and acquaintances previously established in the United States through a porous and poorly invigilated border. And for Porfirian authorities, the Chinese provided the cheap labor needed for their modernization projects.

The crown jewel of the modernization drive was the Tehuantepec railway and its terminal ports in Coatzacoalcos and Salina Cruz. These enabled Mexican politicians to accomplish a long-sought dream of opening an interoceanic route across the country's narrowest stretch. Thousands of Cantonese laborers participated in the construction and, paradoxically, it was in the port that they helped build that Mexican authorities would eventually shut the doors to them. Once finished, the route became the continent's fourth interoceanic passage and Salina Cruz the largest and most modern port in the Mexican Pacific. As such, it became an integral part of the transpacific maritime network and an in-between place where different international actors negotiated and disputed their interests.

Together with Salina Cruz, CCSC's vessels became the loci where all the aforementioned interests not only converged but also collided. As thousands of Cantonese arrived in Mexico, with many of them prospering economically and becoming small merchants, antagonism against them grew in a country where the authorities had long labeled them as inassimilable and incompatible with the ideal national

Mestizo prototype. Besides the usual racialized and racist arguments discussed in previous chapters, the *Suisang* case revealed another powerful type of exclusionary discourse, one that disguised ideology and personal interests in medical terminology. This type of narrative was used not only by Mexican politicians and journalists, when they accused Chinese of polluting the body of the nation with their unsanitary conditions and their diseases, but also by CCSC's doctors, owners, and representatives, who used the medical jargon to privilege their own economic interests over the wellbeing of the passengers who provided them with most of their income. In effect, both parties became entangled in a fruitless technical discussion about an eye disease that masked their actual intentions.

In addition, the *Suisang* controversy and its contextualization exposed a fundamental contradiction held by Porfirian elites who, on the one hand, saw Chinese as cheap, exploitable labor useful for their economic objectives. On the other hand, they complained about the poor and unsanitary conditions of Chinese steerage passengers, without reflecting on the fact that this was directly related to the precarious salaries that Mexicans were willing to pay them. This double vision served these elites to support the arrival of Chinese at some moments and deter it at others. Finally, the Mexican authorities clearly made a distinction in terms of class, not only by allowing the ill second-class passengers to enter the country while rejecting those from steerage, but also by subjecting the healthy steerage travelers to intrusive and scrutinizing hygiene measures that the former never had to face.

The uncertain trajectory of the CCSC claim, which could not be settled due to the decline of the Díaz government and the outbreak of civil war, portended the future for the Pacific strategy put in place during the Porfirian regime. As we will see in the following chapter, the outbreak of the Mexican Revolution and the First World War, combined with the transcendental inauguration of the Panama Canal and the appearance of new technologies, would again transform the relationship of the Mexican coast with the transpacific world.

6
Ancon, 1914
*Revolutions and the End of the Porfirian Transpacific System**

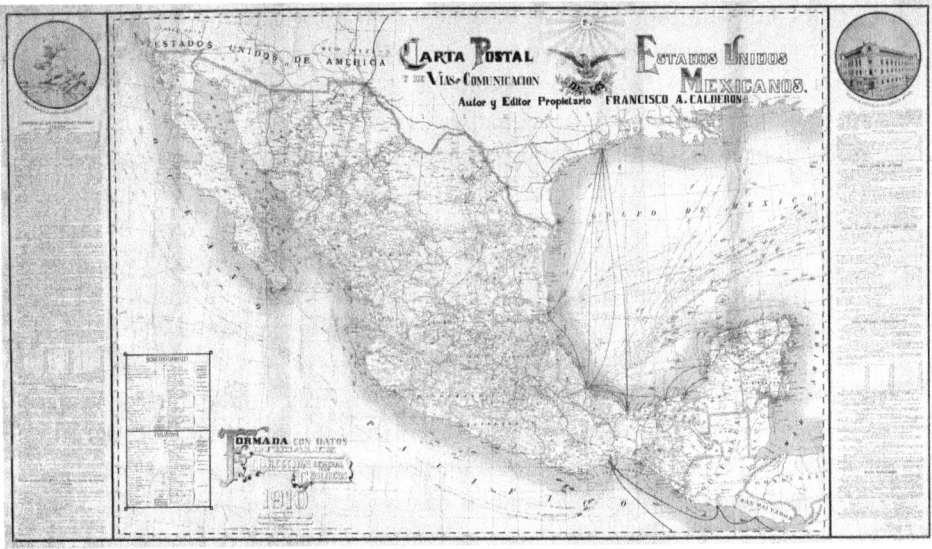

Map 6
Francisco A. Calderón, *Carta postal y de vías de comunicación de los Estados Unidos Mexicanos* [Mexican postal chart and communication routes]. Map. Mexico City: Dirección General de Correos, 1910. From Mapoteca "Manuel Orozco y Berra," República Mexicana 7. Accessed December 15, 2023. https://mapoteca.siap.gob.mx/cgf-rm-m27-v7-0333/.

---

* Data from some maritime companies and travelers appeared in my contribution to the book *Intercambios, actores, enfoques. Pasajes de la historia latinoamericana en una perspectiva global*. I thank editor Aarón Grageda for permission to reuse.

In the early morning of Saturday, August 15, 1914, hundreds of people gathered in the port of Cristobal, Panama, to witness the departure of *Ancon*, a 9,332-ton Panama Railroad Company steamer. While it usually transported construction materials, today it carried a large contingent of prominent local personalities, including Belisario Porras, the president of Panama, his wife, ministers in his cabinet, as well as dozens of diplomats, particularly from the United States. They had arrived in celebration of the official launch of the Panama Canal, the watercourse that made it possible for people and merchandise to travel directly between the Atlantic and the Pacific without having either to use the long and dangerous Cape Horn route or to transfer to any transcontinental railroad.[1] *Ancon*'s movement from Cristobal in the Caribbean to Balboa in the Pacific lasted nine hours and forty minutes. Its traverse was widely covered by the Panamanian press and cheered throughout.[2]

The inauguration was not met with the same widespread enthusiasm in other parts of the world, however. In Europe, the news was overshadowed by the coverage of the Great War, which had started only two weeks earlier. In Mexico many rightfully saw it as a serious challenge to the success of the transcontinental Tehuantepec railroad and its adjacent ports of Salina Cruz and Puerto Mexico, which had been capturing increasing amounts of transoceanic traffic since 1907. Even more threatening for Mexican interests was that the country was in the midst of a civil war that had followed the 1911 deposition and exile of Porfirio Díaz. Given this context, the Panama Canal not only offered a safer and easier transoceanic passage but would also become the route mandated for use by US steamers and companies, which comprised the bulk of Tehuantepec's users. Its owner and operator was the US government, after all.

This chapter explains how a series of military, social, and technological revolutions occurring during the 1910s—the Mexican civil war, World War I, the opening of the Panama Canal, and the introduction of oil technologies in navigation—transformed Mexico's transpacific connections as they had been conceived during Porfirian times. It begins with an overview of how the country's Pacific coastline had developed by the end of the Porfiriato and then explains how and why this scenario began to change. With its participation in the construction and inauguration of the watercourse and later in the US war efforts, *Ancon* symbolizes the beginning of the end of Mexico's transpacific world as it had evolved during the three and a half decades of Porfirian administration.

---

1. The trip from New York to San Francisco was reduced to 8,370 kilometers, compared to 20,900 kilometers via Cape Horn. Marine & Oceans, "The Panama Canal Turns 109: Interesting Facts about This Strategic Location," last modified September 8, 2023, https://marine-oceans.com/en/the-panama-canal-turns-109-interesting-facts-about-this-strategic-location/.
2. Ralph Eldin Minger, "Panama, the Canal Zone, and Titular Sovereignty," *The Western Political Quarterly* 14, no. 2 (June 1961): 544–54. See also David McCullough, *The Path between the Seas: The Creation of the Panama Canal, 1870–1914* (New York: Simon and Schuster, 1977).

Mexico's Pacific Connections and Circulation of People at the End of the Porfiriato

By 1910, and after three decades in power, Porfirian elites had aged. Several longtime key cabinet members had died,[3] leaving decision-making ever more centralized. That summer, Díaz won the election for the eighth time and did not seem overly concerned with the fallout from the opposition's allegations of fraud.[4] Instead, his team focused on preparing an ostentatious set of festivities to celebrate both his eightieth birthday and the country's centennial, which, by a mixture of chance and clever nepotism, landed on the same date: September 15.[5]

Mexico had grown significantly and had diversified its Pacific links since the administration's early days. Looking seaward, the nation's coastal populations went from waiting on an irregular flotilla of steamers and sailboats to counting on daily steamship connections throughout its ports. They could now rely on multiple weekly runs to and from the United States, notably to San Francisco, as well as to Central America, extending all the way south to Panama. There were also regular trips to Hawai'i plus a monthly Asia-bound sailing aboard the China Commercial Steamship Company (CCSC) vessels, discussed in Chapter 5. Additionally, Japan's Toyo Kisen Kaisha (TKK) wanted in on the action, having already begun negotiations for a second direct transpacific route.[6] National land connections had improved as well. Coastal communities went from depending on the irregular circulation of horse and buggies, which were often impeded by weather, difficult dirt roads, and mountainous topography, to having four major railway-serviced ports (Guaymas, Mazatlan, Manzanillo, and Salina Cruz), with a fifth (San Benito) with rail access just some thirty kilometers away. A regular system of stagecoaches operated throughout.[7] Furthermore, the Tehuantepec railroad connected the two coasts with some sixteen daily trains, making it one of the shortest and fastest ways to cross the continent.[8]

---

3. Two that have been mentioned in this work are Matías Romero, who held different posts, most notably Mexico's representative in Washington in the last decades of the nineteenth century, and Ignacio Mariscal, Díaz's secretary of foreign affairs from 1885 until his death in April 1910. For a discussion of the marking of their deaths and the importance of state funerals to Porfirian rule, see Matthew D. Esposito, *Funerals, Festivals, and Cultural Politics in Porfirian Mexico* (Albuquerque: University of New Mexico Press, 2010).
4. This topic will be discussed in the following section.
5. Mexico's War of Independence was supposed to have started in the early morning of September 16, yet, since Porfirio Díaz's birthday was on September 15, he moved the commemoration to the previous night so that the dates coincided.
6. The case of TKK will be further explained later in this section.
7. See map in Paul Garner, *Porfirio Díaz* (London: Pearson, 2001), 256–57.
8. The following source sustains that there were up to sixty daily trains in Tehuantepec by the end of the Porfiriato: Nemesio Rodríguez, *Istmo de Tehuantepec: de lo regional a la globalización o apuntes para pensar un quehacer)* (Oaxaca: Gobierno del Estado de Oaxaca, 2003), 3. Roxana Arce Ibarra, *Los transportes en el istmo de Tehuantepec* (Mexico City: UNAM, 1949), 147, finds that number exaggerated and rather talks about sixteen trains.

In terms of its maritime policy, the Porfirian government went from providing costly subventions to offering tax reductions and exemptions to foster international connections. The latter could include a decrease of 20–60 percent on tonnage and sanitary dues as well as up to a 100 percent exemption from municipal and other federal taxes. By 1910, four Pacific coast companies received these types of benefits: the Mexican-owned Compañía de Vapores de la Costa del Pacífico, circulating between San Francisco, California, and the northern Mexican ports of Ensenada, San José del Cabo, La Paz, Guaymas, Mazatlan, Altata, Santa Rosalía, and Bahía Magdalena;[9] CCSC,[10] which, as we have seen, linked Hong Kong with Manzanillo or Salina Cruz in a monthly trip; the German Kosmos, whose vessels departed Hamburg and landed in several European and South and Central American ports before reaching the Mexican coast—notably Salina Cruz, Acapulco, Manzanillo, and Mazatlan—before continuing on to San Francisco;[11] and finally the US-based Pacific Mail Steamship Company (PMSS), discussed in Chapter 1, which transited between San Francisco and Panama, stopping at San Blas, Manzanillo, Acapulco, and other Central American ports. All these companies would also dock elsewhere when and where demand was sufficient.[12]

By the time of the centennial, the Mexican authorities only rarely reverted to a policy of subventions, and then only when they believed a region with enough economic potential could benefit from having additional international maritime connections.[13] Three Pacific-based companies received this kind of support. Compañía Naviera del Pacífico received $2,333.33 per round trip between San Diego, California, and Bahía Magdalena and Ensenada, in Baja California;[14] M. Jebsen Line, circulating between Seattle, Washington, and Corinto, Nicaragua, received a $5,000 monthly stipend for stopping at Manzanillo and Salina Cruz; and Compañía Oriental de Navegación or TKK, which, as we will see below, negotiated $10,000—limited to $120,000 annually—in subsidies per round trip between Yokohama or Kobe and Manzanillo and Salina Cruz.[15]

---

9. "Las Compañías Navieras y sus obligaciones según contratos," *El Correo de Chihuahua*, February 14, 1907, 2.
10. Sometimes referred to as the Eng Hok Fok Co., in reference to the first president of the company.
11. Kosmos circulated along the Mexican coast from 1899. It had a monthly route, stopping at dozens of ports between Hamburg and San Francisco, so its vessels often ran late. They carried mainly cargo but always had room for at least fifty passengers. "Nueva línea de vapores," *El Economista Mexicano* 28 (August 1899–January 1900): 123. In late 1910, Kosmos announced its withdrawal from docking at Salina Cruz due to the higher priority given to larger vessels from other companies, but it continued with its other stops in the Mexican Pacific. "La línea Kosmos," *Diario del Pacífico*, October 20, 1910, 2.
12. Archivo Histórico Genaro Estrada de la Secretaría de Relaciones Exteriores (AHGE-SRE), L-E-187, "Cuarta Conferencia Internacional Panamericana," 1910, 59.
13. Sandra Kuntz Ficker, *El comercio exterior de México en la era del capitalismo liberal, 1870–1929* (Mexico City: El Colegio de México, 2007), 106.
14. This company had several other national routes, which are not mentioned here because our focus is the maritime links with ports situated outside Mexico.
15. AHGE-SRE, L-E-187, 55. In the early 1900s, the Pacific-facing companies that received subventions from the Mexican government were PMSS, Compañía de Vapores de la Costa del Pacífico and Compañía del Desarrollo de Baja California. "Establecimiento de líneas de vapores que favorezcan el tráfico mercantil," *El Economista Mexicano* 33 (October 1901–March 1902): 196.

During the first decade of the twentieth century, a number of other international companies connected with Mexico's west coast without receiving benefits from the government.[16] These were Compañía del Desarrollo de Baja California, which made two monthly trips between San Diego, California, and several of Baja California's ports, including Ensenada, San Quintín, and Isla de Cedros;[17] Canadian Mexican Pacific Steamship Line, which navigated from Victoria and Vancouver, British Columbia, to Guaymas, Mazatlan, San Blas, Manzanillo, Acapulco, and Salina Cruz;[18] New Fast Steamer, renamed Salvador Railway Company Steamship Service, traveling between El Salvador and Salina Cruz;[19] and American–Hawaiian Steamship Company (AHSC), sailing between Hawai'i and Salina Cruz.[20]

The active circulation of Chinese goods and peoples enhanced the movement of all these vessels. The port of Mazatlan, Sinaloa, offers a case in point. The 1910 census listed 663 Chinese in Sinaloa. Many owned retail businesses in Mazatlan, such as Yuen Fo San & Co., Ramon Kooc & Co., and Hop, Ley & Co., each with an initial capital of Mx$2,000; Fon Chon Fay & Co., with Mx$4,750; Leon & Co., with Mx$1,800; and Hon Yuen & Co., with Mx$3,000.[21] The *Diario del Pacífico*'s shipping reports during the summer of 1910 show that Chinese were, after Mexicans, the most mobile nationality, not only in terms of people but also merchandise sent and received. For instance, in the month of July, Mazatlan's Chinese community registered the following. On July 5, Ramon Kooc received ten sacks of rice and forty-six

---

16. The exact dates when these companies' ships circulated are uncertain since they either did not sign a contract with the government or the contract was not found. It is also possible that some of these companies received some governmental aid before 1910. Yet none of them were mentioned in the official document on maritime communications in the Pacific, presented by the Mexican government at the Fourth Pan-American Conference: AHGE-SRE, L-E-187.
17. Sergio Armando Gallegos López, "Los fletes marítimos: factor de desarrollo en el comercio exterior de México" (BA thesis, Universidad Nacional Autónoma de México/FCPyS, 1972), 11. This source also mentions two more companies that traveled abroad, those belonging to Juan B. Abaroa and Luis H. Martínez. Yet the following primary sources cite them as traveling only within Baja California: all the 1909 issues of Dirección General de Correos, *Movimiento Probable de Vapores*; "Las Compañías Navieras y sus obligaciones según contratos," *El Correo de Chihuahua*, February 14, 1907, 2. The latter added that Compañía del Desarrollo de Baja California had a contract with the government in 1907 to transport military and governmental employees as well as colonizers brought in by the Ministry of Public Works for half the normal price.
18. "Algo de las costas mexicanas," *El Correo de Chihuahua*, February 17, 1908, 1. The company is also listed in the 1909 issues of the Dirección General de Correos's *Movimiento Probable de Vapores*, available for consultation in the Biblioteca Miguel Lerdo de Tejada, and of the *Guía Oficial de Ferrocarriles y Vapores Mexicanos*, available for consultation in Hemeroteca Nacional's Fondo Reservado.
19. The first name appears only in the 1912 issues of the *Guía Oficial de Ferrocarriles y Vapores Mexicanos*. Then it gets substituted by the second name as of the January 1913 issue of the same *Guía*.
20. Compañía de la Guía Oficial S.A., *Guía Oficial de Ferrocarriles y Vapores Mexicanos*, February 1913. The case of AHSC is peculiar in the sense that at the height of the Tehuantepec railroad, a third of its stock was owned by the railroad proprietors, that is, the Mexican government and Weetman Pearson.
21. Rigoberto Arturo Román Alarcón, *La economía del sur de Sinaloa, 1910–1950* (Mazatlan: Instituto Municipal de Cultura, Turismo y Arte de Mazatlán/DIFOCUR, 2006), 131. Other retail businesses listed in the *International Chinese Business Directory of the World for the Year 1913* include those of Chong Yuen Lee, Fong Hie Gui, Fong Sang Wo, Hop Lee, Hop Wo, Juan Pat, Kwong Qui, Quong Sang Yuen, Quong Wah Shing, Quong Wo, San Wo, Ton San Foy, Wah Ling, and Wah Lung Jan.

of generic goods from Manzanillo.[22] On July 7, the German steamer *Sisak*, coming from Hamburg, disembarked a barrel of glassware for Miguel Wongpek; ninety-two packages that included newspaper, cinnamon, dry fruits, tinned food, pastas, beans, oil, sugar, starch, potatoes, and wooden artifacts for Juan Pat; and thirty-six packages of fish and shrimp for Ramon Kooc. On July 9, Yuen Kui received twelve packs of hats, clothes, shoes, fabric, and liquors from Guaymas, and two out of the nine passengers landing from San Francisco aboard *Newport* were Chinese. On July 10, the steamer *Manuel Herrerías* brought nine packages of cheese and skins for Juan Pat from Ensenada and then departed for San Diego with eight Chinese on board. On July 12, *Rio Yaqui* brought from Teacapan twenty-two sacks of corn destined for Jose Wongpek. The next day, fifty-three Chinese arrived from Manzanillo aboard *Luella*. On July 15, Ramon Kooc exported thirty-six packs of fish and shrimp worth Mx$440 to San Francisco, and the steamer also carried a single Chinese person. On July 16, the Norwegian steamer *Transit*, coming from Salina Cruz, brought seventy-three Chinese and miscellanea for Miguel Wongpec and the Yacho brothers.[23] On July 22, Ton On Hing and Tung Sang Wo received mail. A day later, the steamer *San Juan* from San Francisco brought ten Chinese passengers as well as silks and porcelain for Chon Yuen. The next day it departed with six packs of fish and skins worth Mx$220 from Juan Jiho.[24] A similar pattern was repeated in the successive months with the following Mazatlan businessmen as the most active in sending and receiving merchandise: Chon Yuen Lee, Hong Tac, Hong Yuen, Hop Ley, Juan Chan, San Wo, Tong Chong Tay, Tong San Fo, Tung Sang Wo, Wing Gung, the Yacho brothers, Yuen Chan, Yuen Fo San, Juan Jiho, Juan Pac, and Ramon Kooc—the last three formed part of the board of directors of the port's main Chinese organization, the Asiatic Club. All these exchanges reveal that the wide variety of products offered by the Chinese retail stores relied on community networks that were mostly North American but also included China and Europe.

The expansion of Mexico's maritime connections was undertaken hand in hand with the improvement of its port infrastructure, which, in turn, was highly influenced by the development of the country's railroad network, built with the participation of Chinese workers.[25] Not surprisingly, by 1910 the four largest international Pacific ports all had direct rail connections. Guaymas, in the state of Sonora, was, in 1882, the first. Its track linked to the border town of Nogales, where the line subsequently joined the United States' railway system. Guaymas remained the area's

---

22. Since Manzanillo was one of the transpacific terminals for TKK's and CCSC's transpacific vessels, some of this merchandise could have come from China.
23. As mentioned above, since Salina Cruz was one of TKK's and CCSC's transpacific terminals, some of this merchandise could have come from China.
24. All this information came from monitoring the section "Por el Muelle," found on page 2 of *Diario del Pacífico* during the entire month of July 1910.
25. Chapters 2, 3, and 5 have discussed Chinese participation in railroad construction.

main port, highly influenced by border zone transactions.²⁶ Manzanillo was next. Its original 1889 rail line, financed by the US-owned Compañía Constructora Nacional Mexicana, solely linked it with the state capital. But in 1908, the section uniting Colima city with Guadalajara—the largest city in the west—was inaugurated by Porfirio Díaz himself. Between these dates Manzanillo experienced an urbanization process that provided it with an upgraded central plaza and cobblestone streets, banks, electricity, and health services. The port, however, remained a highly unsanitary location during the rainy season due to the proximity of a large lagoon that attracted all sorts of insects, many of which were known to be carriers of infectious disease. An 1899 contract with US businessman Edgar K. Smoot triggered the port's expansion, with the building of a seafront, a jetty, a series of piers, and a breakwater, among other infrastructure. While the original project duration was listed at four years, construction continued beyond the centennial.²⁷ Salina Cruz got rail access in 1894 with the establishment of the Tehuantepec transcontinental railroad. From then on, the port and the whole line all the way to the Gulf of Mexico experienced a series of improvements—described in the previous chapter—becoming the country's most modern intermodal region as of 1907. Mazatlan, the last of the Pacific ports boasting a direct railway connection during the Porfiriato, could blame the Sierra Madre's harsh topography for the delay. In fact, the line, finished in 1909, only connected with other Pacific ports and failed to tie it to the central states. In this sense Mazatlan was different from the other three in that its early port infrastructure was linked not to the expansion of the railway but rather to Sinaloa's agricultural, mineral, and, later, industrial boom. The capital from those activities financed port improvements in the late nineteenth century.²⁸ Manzanillo and Salina Cruz became the only ports with regular direct transpacific traffic precisely due to their improved railway connections, the former with Colima and Guadalajara and

---

26. See Juan José Gracida Romo, "Guaymas, notas para la historia comercial del puerto, 1820–1910," in *Los puertos noroccidentales de México*, ed. Jaime Olveda and Juan Carlos Reyes (Zapopan: El Colegio de Jalisco, 1994), 199–212.
27. Blanca Estela Gutiérrez Grageda and Héctor P. Ochoa Rodríguez, *Las caras del poder: conflicto y sociedad en Colima, 1893–1950* (Colima: Universidad de Colima/Gobierno del Estado de Colima/Conaculta, 1995), 15–18, 35; Héctor P. Ochoa Rodríguez, "Manzanillo, el intrincado despertar de un puerto," in *Los puertos noroccidentales de México*, ed. Jaime Olveda and Juan Carlos Reyes (Zapopan: El Colegio de Jalisco, 1994), 116–22. See also Karina Busto Ibarra, *El Pacífico mexicano y sus transformaciones: integración marítima y terrestre en la configuración de un espacio internacional, 1848–1927* (Mexico City: El Colegio de México, 2022), 336–38, 390–94.
28. Luis Antonio Martínez Peña, "Mazatlán, historia de su vocación comercial, 1823–1910," in *Los puertos noroccidentales de México*, ed. Jaime Olveda and Juan Carlos Reyes (Zapopan: El Colegio de Jalisco, 1994), 157–78. For more on the specific improvements begun at the end of the nineteenth century, see "Informe del Presidente de la República al abrirse el 3er periodo del 19° Congreso de la Unión, 19 de septiembre de 1899," *El Economista Mexicano* 28 (August 1899–January 1900): 90. For an analysis by a local newspaper of the role played by Mazatlan in the Pacific port system as well as the problems faced by the port after the 1909 inauguration of the railway, see "Nuestras comunicaciones marítimas y terrestres," *Diario del Pacífico*, December 6, 1910, 6; "El comercio exterior de Sinaloa," *Diario del Pacífico*, December 27, 1910, 6.

from there to Mexico City and the northern states, and the latter with its easy access to the Atlantic via the transcontinental railroad.

While its Pacific maritime links had grown, there were still limitations. According to the government, in a document presented to the Fourth Panamerican Conference, held in Buenos Aires in August 1910, the most notable weakness was the lack of communications with South America. Even though some contracts had been signed, they were not enforced. This was due to the fact that "companies, on the one hand, lacked commercial incentives to carry [the contracts] out, and the governments, on the other hand, did not encourage them."[29] In 1899, two companies began dispatching steamers from Valparaiso, Chile, to Mazatlan, stopping at various ports along the way. These were the British Pacific Steam Navigation Company and the Chilean Compañía Sudamericana de Vapores.[30] By 1903, both had retired their routes north of Panama.[31] It was not until 1910 that Japan's TKK reconnected Mexican, Peruvian, and Chilean ports.[32] In 1913 the South American Compañía Peruana de Vapores y Dique del Callao signed a contract establishing regular transport between Callao, Peru, and Salina Cruz. In exchange for tax reductions, it agreed to use exclusively the Tehuantepec railroad to transport merchandise across the continent.[33] Yet it is uncertain that this ever materialized. Besides the lack of routes to South America, other important obstacles, according to *El Economista Mexicano*, a biannual journal specializing in economic affairs, included the scarcity of successful Mexican-owned companies that traveled abroad as well as the lack of shipyards that could fabricate large steamers, leaving the country dependent on foreign variables and interests for its maritime connections.[34]

In 1910, two companies offered direct transport from Asia to Mexico: CCSC and TKK.[35] The first was a Chinese venture that started its offerings between Hong Kong, Moji, Kobe, Yokohama, Manzanillo—and later Salina Cruz—in 1903 and was discussed at length in the previous chapter. The second was a Japanese company created in the mid-1890s with a governmental subvention in order to enter the transpacific trade between Yokohama and San Francisco. It did so with three 6,000-ton British steamers—*Nippon Maru*, *America Maru*, and *Hong Kong Maru*—sailing approximately every month. The first entered San Francisco Bay on

---

29. AHGE-SRE, L-E-187, s/f.
30. "Informe del Presidente de la República al abrirse el 3er periodo del 19° Congreso de la Unión el 19 de septiembre de 1899," *El Economista Mexicano* 28 (August 1899–January 1900): 88–92.
31. "Navegación," *El Economista Mexicano* 35 (October 1902–March 1903): 36.
32. The case of this company will be discussed more thoroughly later in this section.
33. "Servicio de vapores entre Salina Cruz y el Callao (Peru)," *El Economista Mexicano* 57 (October 1913–March 1914): 281–83.
34. "Alientos a la marina y a la navegación," *El Economista Mexicano* 57 (October 1913–March 1914): 246. While this diagnosis was written in 1913, when the Mexican Revolution had already started, it applies to the last years of the Porfiriato as well. In effect, of all the companies listed in the previous pages, only three had Mexican capital: Compañía de Vapores de la Costa del Pacífico, Compañía Naviera del Pacífico, and Compañía del Desarrollo de Baja California.
35. Often referred to as Compañía Oriental de Vapores in Spanish and Oriental Steamship Company in English.

January 1899 and became the largest and fastest commercial carrier ever to land there up to then. It had accommodations for 98 first-class, 40 second-class, and 1,000 steerage passengers.[36] In the summer of 1896, the company sent its agent, Mr. Tomioka, to Mexico to review the possibility of starting a regular service. This did not materialize until 1910.[37] By then, TKK had fabricated three new steamers in Nagasaki's Mitsubishi Dockyard & Engine Works—named *Tenyo Maru*, *Chiyo Maru*, and *Shinyo Maru*—and was ready to open new routes.[38] Voyages to Peru and Chile began in 1905 with only two or three trips per year and without necessarily touching Mexican ports.[39] In October 1908, TKK's general manager, Mr. Shiraishi, visited Mexico with his wife and inspected its Pacific ports.[40] By early 1910, after a TKK executive visited the country,[41] they signed a contract with the Mexican government for up to twelve trips per year between Asian ports and Manzanillo and Salina Cruz. TKK would receive $10,000 in subsidies for every completed round trip.[42] On April 15, the Mexico City newspaper *El Imparcial* heralded the arrival of the *Kiyo Maru* in Manzanillo, which inaugurated TKK's regular route to Mexico.[43] The whole itinerary, named the South American Line, linked Hong Kong, Moji, Kobe, Yokohama, Honolulu, Manzanillo, Salina Cruz, Callao, Arica, Iquique, Valparaiso, and Coronel. The return trip included the same stopovers.[44] With the two lines sailing regularly, imports from Asia broadened. For instance, in April 1910, Mexico received $288,850 worth of merchandise from that continent, making it the third most significant import region after the United States and Europe (with $11.2 million and $6.3 million respectively) and surpassing Canada ($244,011), South and

---

36. E. Mowbray Tate, *Transpacific Steam: The Story of Steam Navigation from the Pacific Coast of North America to the Far East and the Antipodes, 1867–1941* (New York: Cornwall Books, 1986): 62.
37. "Nuestras relaciones con el Japón," *El Economista Mexicano* 22 (August 1896–January 1897): 14–15; "La bandera japonesa en nuestros puertos del Pacífico," *La Gaceta Marítima* 14 (August 1896): 228; "Japanese Line to Mexico," *New York Times*, August 22, 1896.
38. The three new steamers were again some of the fastest and largest to sail across the ocean at the time. *Tenyo Maru*, for instance, set a record on June 29, 1908, that took many years to beat. It sailed from Honolulu to San Francisco in four days, eighteen hours, and thirty minutes. In terms of size, they all surpassed 13,000 tons and had accommodations for 275 first-class, 64 second-class, and 800 steerage passengers. They all possessed electric lights, as well as private bathrooms for first-class passengers, an innovation for the time. Tate, *Transpacific Steam*, 64; *Toyo Kisen Kaisha/Línea Oriental de Vapores. Folleto sobre el servicio directo entre Japón, China, México y Sudamérica*, 1909, 20–21.
39. Tate, *Transpacific Steam*, 66; "Toyo Kisen Kaisha," *The Hong Kong Daily Press*, February 12, 1907, 6.
40. "Un huésped distinguido súbdito del Mikado," *El Tiempo. Diario Católico*, October 28, 1908, 2.
41. "Desde Hong Kong hasta Salina Cruz," *El Tiempo. Diario Católico*, April 13, 1910, 1.
42. AHGE-SRE, L-E-187, 55.
43. "El tráfico marítimo entre México y el Japón," *El Imparcial*, April 15, 1910, 7. The following Mazatlan newspaper cited *Bayu Maru* as the steamer that inaugurated the service: "Nueva línea de vapores," *Diario del Pacífico*, December 13, 1910, 6.
44. Dirección General de Correos, *Noticia del Movimiento Probable de Vapores durante el mes de enero de 1911* (México City: La Ilustración, 1911): 10; *Toyo Kisen Kaisha*, 3. The brochure, written in Spanish, cites *Hong Kong Maru*, *Kiyo Maru*, and *Buyo Maru* as the vessels devoted to the South American Line, and *Tenyo Maru*, *Chiyo Maru*, *America Maru*, and *Nippon Maru* as those servicing the Hong Kong–San Francisco run. Both TKK and CCSC had fixed itineraries and destinations but could stop at additional ports if there was enough demand.

Central America ($80,007 and $14,411 respectively), and Africa ($23,223). Exports to Asia, meanwhile, remained trivial, with only $200 registered for that month.[45] In order to redress the trade imbalance, in October 1910 Mexican entrepreneurs began organizing a business trip to Japan.[46]

The main business for both CCSC and TKK, at least prior to the end of the Porfiriato, was the transportation of Asians—mostly Cantonese—in steerage, many of whom stayed in Mexico.[47] In 1910, over 13,000 Chinese lived in Mexico, mostly in Sonora and the neighboring northwestern states.[48] Since the beginning of the century, thousands of Japanese had also landed in Mexico to work on haciendas—notably of sugarcane—in mines, and in the construction of railroads.[49] Just like their Chinese peers, most were young males between fifteen and thirty years of age who established themselves in the northwestern states, from where a majority eventually crossed into the United States. In contrast to the Chinese, Japanese laborers were never banned from entering the US. Yet by the turn of the century they faced increasing discrimination, particularly in California. Consequently, in 1907 the government of Japan informally agreed to stop the migration of new laborers in exchange for a promise by the US authorities that they would respect the rights of those already living inside their territory.[50] It is calculated that at least 10,000 Japanese arrived in Mexico during the Porfiriato. The majority came from the island of Honshu, particularly from the prefectures of Hiroshima, Wakayama, Yamaguchi, Shizuoka, and Nagano. Another large group came from Fukuoka, on Kyushu Island, and, to a lesser but still significant extent, from Okinawa.[51] While many crossed in CCSC and TKK steamers, others sailed in vessels chartered by immigration companies and/or Mexican businessmen who hired them, often under precarious conditions.[52] This was also the case for a group of over 1,000 Koreans who traveled

---

45. "Aumento en las importaciones," *Diario del Pacífico*, July 13, 1910, 3.
46. "Una excursión al Japón," *Diario del Pacífico*, October 16, 1910, 3; "Relaciones comerciales El Japón y México," *Diario del Pacífico*, October 17, 1910, 1.
47. Chapters 2 and 5 discuss the transportation of Chinese laborers to Mexico.
48. Evelyn Hu-DeHart, "On Coolies and Shopkeepers: The Chinese as *Huagong* (Laborers) and *Huashang* Merchants) in Latin America/Caribbean," in *Displacements and Diasporas, Asians in the Americas*, ed. Wanni W. Anderson and Robert G. Lee (New Brunswick, NJ: Rutgers University Press, 2005), 89; Robert Chao Romero, *The Chinese in Mexico, 1882–1940* (Tucson: University of Arizona Press, 2010), 29, 53.
49. See Chapter 4 for more information on this subject.
50. The pact is often referred to as the Gentlemen's Agreement.
51. María Elena Ota Mishima, "Características sociales y económicas de los migrantes japoneses en México," in *Destino México: Un estudio de las migraciones asiáticas a México, siglos XIX y XX*, ed. María Elena Ota Mishima (Mexico City: El Colegio de México, 1997), 56–63; Tate, *Transpacific Steam*, 231.
52. See, for instance, the cases of the Japanese hired to work in the mines of Boleo, in Baja California, as well as in Las Esperanzas, Coahuila, in 1906 in AHGE-SRE, 9-1-79, "Inmigrantes japoneses. Demanda presentada en Tokio contra la Cia. Oriental de Emigración japonesa de las Minas del Boleo en Santa Rosalía, Baja California por—," 1906; AHGE-SRE, 15-16-65, "Inmigrantes japoneses. La legación japonesa solicita los oficios de esta Sria. para que se haga averiguación acerca del mal trato que se dice reciben," 1906. For more detailed information on Japanese contract workers in Mexico, see Enrique Cortés, *Relaciones entre México y Japón durante el Porfiriato* (Mexico City: SRE, 1980), 84–102; María Elena Ota Mishima, *Siete migraciones japonesas en México, 1890–1978* (Mexico City: El Colegio de México, 1982), 51–62.

aboard the chartered British steamer *Ilford* from the port of Inchon to Salina Cruz in the spring of 1905. They all ended up working under strenuous conditions in Yucatán's henequen plantations.[53]

By 1910 most Asians living in Mexico had transitioned from laborers to small merchants, primarily engaged in retail, although they also prospered in other businesses such as restaurants, bakeries, and laundries.[54] Sonora represents a case in point as that border state was home to the country's largest concentration of Chinese. In effect, of the 13,203 Chinese registered in the 1910 census, roughly a third—or 4,486—lived there, where they composed the largest foreign minority, surpassing US nationals, the second largest group, by over 1,000 people. Only thirty-seven were women, confirming the pattern of an overwhelmingly male population. Most were small entrepreneurs who had succeeded at establishing prosperous *abarrotes* or small grocery stores and had slowly overtaken Europeans as the region's main providers, present in all state districts to the point that they held a virtual retail monopoly. They maintained competitive prices, offered a wide variety of products, and peddled to isolated communities that had never been serviced. The owners usually employed relatives and friends brought from their hometowns who, after working for several years, would amass enough earnings to establish their own *abarrotes*, hiring their own acquaintances and therefore reproducing the pattern over and over.[55]

One of the main reasons for their prosperity had to do with interconnectivity with other Chinese capitalists throughout the area. While the main relationships were with those living in California and southeastern China, those in Mexico could easily locate others of their fellow countrymen and carry out transactions with them as well. This becomes evident when reviewing the International Chinese Business Directory of the World from the time. This document updated the names and addresses, in both Chinese and English, of thousands of businesses and associations in hundreds of cities in China, Japan, India, Indochina, the Malay peninsula, Siam, Java, Sumatra, Borneo, Australia, Africa, New Zealand, Hawaiʻi and continental United States, the Philippines, Canada, the West Indies, and Latin America. In the Mexican section, some 600 establishments appeared—almost half in Sonora. Networking was made easy: Each locality offered a list of merchants with concise contact information.[56] From their many connections, Chinese in Mexico could

---

53. Alfredo Romero Castilla, "Huellas del paso de los inmigrantes coreanos en tierras de Yucatán y su dispersión por el territorio mexicano," in *Destino México: Un estudio de las migraciones asiáticas a México, siglos XIX y XX*, ed. María Elena Ota Mishima (Mexico City: El Colegio de México, 1997), 123–66. See also Wayne Patterson, *The Korean Frontier in America: Immigration to Hawaii, 1896–1910* (Honolulu: University of Hawaiʻi Press, 1988), 146–47.
54. To learn more about the wide range of occupations of the Chinese living in Mexico in the first decades of the twentieth century, see Chao, *Chinese in Mexico*, 98–118.
55. Hu-DeHart, "Coolies and Shopkeepers," 89–94; Chao, *Chinese in Mexico*, 58.
56. *International Chinese Business Directory of the World for the Year 1913* (San Francisco: International Chinese Business Directory Co., 1913), 1569–88.

often obtain cheaper and varied supplies as well as credit with no or low rates of interest; they could exchange stock and know-how; and, when partaking in a joint transnational venture, they could receive regular dividends or even monthly salaries ranging from $30 to $75. Working together, merchants could obtain yearly profits of $1,800 to $3,000.[57]

The Japanese were also moving around throughout the Pacific. In 1910 the most publicized arrival was that of Shintaro Morimoto, president of TKK, who landed in mid-February "with the purpose of carefully studying maritime and commercial matters in Mexico in order to contribute to strengthen the commercial relations" between the countries.[58] He first visited Mexico City, remaining for two months. In mid-April he traveled to Manzanillo with N. Kobayashi, a member of the Japanese legation, where they embarked on TKK's *Kiyo Maru* to Salina Cruz to examine the Tehuantepec railroad. Before sailing south of the border to continue his international study, Morimoto then ventured to Chiapas,[59] meeting with the country's first Japanese colony, then composed of close to one hundred small entrepreneurs, many of whom, after arriving as colonists and farmers, had transitioned to small merchants and owned various retail establishments.[60]

Beyond steerage passengers and businessmen, the sailors themselves were another important group of transpacific travelers. Since PMSS began passenger transport in 1867,[61] it had created an industry-wide hiring pattern: English-speaking, white officers commanded largely Chinese crews. Companies hired the former to cater to their first-class, Westernized clientele, who were used to seeing whites in power, while the latter were allegedly more obedient, skilled, and, overall, up to 50 percent cheaper than their white counterparts in lower-paid positions. TKK hired its first and second officers from Europe or the United States and recruited its third and fourth officers and crews from Japan rather than China.[62] A substantial number of Mexicans crewed the ships to the point that, in 1909, the country's consul in Hong Kong urged the minister of foreign affairs to take measures to stop Mexican sailors from departing without "the proper documents that prove or justify their

---

57. Chao, *Chinese in Mexico*, 97–98, 124–25. Chao identified four different types of Chinese merchants living in Mexico at the time: the immigrant merchant magnate, the medium-sized merchant, the sole proprietor, and the small merchant, the latter being the most numerous. For the characteristics and specific examples of each, see Chao, *Chinese in Mexico*, 118–29.
58. "Un enviado da compañía de vapores Kisen Kaisha," *El Imparcial*, February 20, 1910, 4. In this article, Morimoto is described as a TKK representative, but in a later article in the same newspaper as well as in "Llega millonario japonés," *El Correo de Chihuahua*, February 16, 1910, 1, he is referred to as the president of the company.
59. "Un enviado de la compañía de vapores Kisen Kaisha," *El Imparcial*, February 20, 1910, 4; "El tráfico marítimo entre México y el Japón," *El Imparcial*, April 15, 1910, 7; "Desde Hong Kong hasta Salina Cruz," *El Tiempo. Diario Católico*, April 13, 1910, 1.
60. Ota, *Siete migraciones*, 46–49. Cortés, *Relaciones*, 80–81; *Relación de la visita oficial a la zona de la colonia Enomoto de la Chiapas, sur de México* (Mexico City: 1958), 56–61. See Chapter 4 for more information on the Japanese in Chiapas and for other Japanese businessmen visiting Mexico at the turn of the century.
61. See Chapter 1.
62. Tate, *Transpacific Steam*, 238–40.

nationality . . . abroad." He complained about the increasing number of Mexican sailors in Hong Kong who found themselves jobless, penniless, paperless, unable to speak the language, and therefore with no possibility of negotiating a contract to take them back home. He claimed to have helped "some of those who seemed to deserve aid" by getting them a "job in the ships and to some others [he] had managed to make . . . the steamers that do the service . . . to Manzanillo or Salina Cruz to take them back to Mexico, working on board [in exchange] for their fare and food."[63] The ministry took action by circulating an order to coastal authorities not to

> allow the embarkation of Mexican sailors under any circumstance aboard foreign vessels without [them] being provided with the documents that prove their citizenship, . . . [as well as with] contracts that explicitly oblige to their repatriation. . . . The harbormasters must [also] make certain that . . . all sailors hired in their port return to it, and if there is one missing, they must find out why . . . [and prove] it with a written evidence of desertion, if that was the case, . . . or . . . with the receipt that shows the severance paid [to the sailor].[64]

While it is unlikely that local bureaucracies could gather such written evidence, the fact that the ministry took the issue seriously suggests that the number of Mexican sailors was not insignificant.

State envoys made frequent journeys during this time. Of these, the December 1910 arrival of the imposing Japanese-made war steamers *Asawa* and *Kasagi* made a splash in Manzanillo and in the nation's press. The former housed 900 sailors and 30 officers while the latter carried 560 crew members and 32 officials. The stopover was just a slice of an ambitious government-sponsored expedition to salute the nations of the Pacific Rim by visiting their main ports. The envoys had arrived via San Francisco where, according to some reports, the local aristocracy had been disinterested and dismissive, even in social events held in their honor. In contrast, they received a magnificent welcome both in Manzanillo and during their brief Mexico City foray, where Porfirio Díaz greeted them personally.[65] They returned to the coast by railway and proceeded to Acapulco, where they loaded enough coal to continue to Salina Cruz, the last stop in Mexican waters before sailing towards Panama.[66]

Transpacific movement was also common among Mexican, Chinese, and Japanese diplomats during this time. José Martín Rascón pioneered the trend, opening the nation's first legation in Tokyo—and Asia—in 1891; poor health forced

---

63. AHGE-SRE, 18-25-83, "Mexicanos en el extranjero sin documento de nacionalidad. Informa el cónsul de Hong Kong," November 17, 1909, s/f.
64. AHGE-SRE, 18-25-83, January 27, 1910.
65. Cortés, *Relaciones*, 119.
66. "Vendrán los cadetes del imperio japonés. Llegarán el próximo verano," *El Imparcial*, April 15, 1910, 1; "Llegan los barcos nipones," *Diario del Pacífico*, December 7, 1910; "Arribaron el Asawa y el Kasagi," *Diario del Pacífico*, December 16, 1910, 1; "Interviu con un marino japonés," *Diario del Pacífico*, December 17, 1910, 1; "Los marinos nipones," *Diario del Pacífico*, December 21, 1910, 6; "Los marinos del Japón en México," *Diario del Pacífico*, December 24, 1910, 1.

him to resign his position after only fourteen months. Luis G. Pardo administered the office from 1893 to 1895, when Mauricio Wollheim took charge. During his years abroad, Wollheim traveled widely, as described in Chapter 4. In 1898 he sent the corpse of the Mexican consul in Yokohama, Eduardo J. Plaza, aboard a PMSS steamship for repatriation. Just two months later, on December 10, 1898, he boarded PMSS's San Francisco-bound steamer *China* at the end of his posting.[67] Ramón G. Pacheco and the Cuban-born lawyer Carlos Américo Lera followed in quick succession in the Tokyo office, while Fidel Rodríguez Parra took charge of the Yokohama consulate (which was soon moved to Kobe). Pacheco returned to the Tokyo post in 1907 and remained there until the end of the Porfiriato.[68] In China, while Pablo Herrera de Huerta opened the first Mexican legation in Peking in 1903, it was not until the following year that, finding himself in Asia once again, Mauricio Wollheim became the country's first official representative there. In February 1905, deteriorating health forced his retirement and departure. On his way back to Mexico he stopped in Yokohama and, on April 15, he left Asia for good aboard US steamer *Coptic*, sailing to San Francisco. He then continued to Mexico City by rail.[69] Between 1905 and 1910 Carlos Américo Lera, Ignacio Altamira (who died in Peking), Ramón Pacheco, Leopoldo Blázquez, and Alfonso Acosta successively occupied the post. Their Chinese counterparts in Mexico City between 1904 and 1910 were Liang-Cheng, Hsun Liang, Tan Poishing, Li Ching Hsu, and Shan Chi.[70] Another important member of the legation was Tam Pui-Shum, who spent a decade writing the first Chinese–Spanish dictionary in Mexico, completing it in the summer of 1910.[71] Japan's Mexico City–based ministers and general consuls between 1891 and 1910 included Gozo Tateno, Shinishiro Kurino, Toshiro Fudyita, Jisashi Shinamura, Yoshibuni Murota, Aimaro Sato, Keichi Ito, Koichi Suguimura, Minozi Arakawa, and Konaichi Joriguchi.[72] All these diplomats brought co-workers and family members with them and often traveled back and forth, enhancing the need for and number of transpacific crossings.

In the fall of 1910, more diplomats and visitors than ever arrived to attend Mexico's flamboyant centennial celebrations commemorating the hundredth anniversary of the initiation of the War of Independence against Spain. In effect, between July and December 1910, over 53,000 people entered the country, two-thirds of

---

67. AHGE-SRE, L-E-1856, 257.
68. Cortés, *Relaciones*, 103–4.
69. AHGE-SRE, L-E-1856, 352, 401, 404.
70. AHGE-SRE, Lista Diplomática de México en China, Tomo 5.
71. "El diccionario español está traducido al chino," *Diario del Pacífico*, June 7, 1910, 1.
72. Cortés, *Relaciones*, 110–14. Some of them appeared in Chapter 4.

whom were male, mainly from Mexico, the United States, Spain, China, and Great Britain (in that order).[73] Thirty countries sent representatives to the festivities.[74]

Besides taking part in the celebrations, many foreign legations hosted their own parties, exhibits, and presentations of ostentatious gifts and monuments.[75] Event planning had begun in 1907, when a special federal committee and hundreds of local commissions were appointed to put together the month-long series of festivities. During that September the capital's main avenues were decorated with independence-related motifs and electric lights. The city inaugurated monuments, public buildings, speeches, exhibits, parks, and academic congresses. Parades, military bands, and allegorical carriages marched through the city's streets. Public works renovated museums, schools, and numerous other public places. The city's printing presses released many publications, banners, and a special collection of stamps. Celebrations peaked on September 15, the supposed date when the War of Independence had started. In reality, though, the war had begun at dawn on September 16; Díaz moved the commemoration forward to coincide with his eightieth birthday.[76] As a consequence, on September 15, the Gran Desfile Histórico (Great Historical Parade) took to the streets with hundreds of actors in costume portraying peak moments of official history. Allegorical carriages depicting the regime's achievements followed. Some 200,000 people witnessed the event. Meanwhile, Porfirio Díaz celebrated an exclusive birthday party in the nearby Chapultepec Castle with some 15,000 invited guests, mostly political and economic elites as well as the foreign diplomatic corps. Finally, at dusk everyone gathered in the central plaza or *Zócalo* (the masses at street level, the elites on the balconies of

---

73. The complete figures for the economic year from July 1, 1910, to June 30, 1911, are the following: 79,484 people registered their entrance into Mexico; 61,073 were male and 18,412 were female. Of them, 25,747 entered through the northern border with the US, 18,006 via the Gulf of Mexico, and 7,179 via Pacific ports, while only 30 did so by land from Guatemala. Of those entering, 26,004 reported to be Mexican, 23,598 from the United States, 5,734 from Spain, 3,959 from China, and the remainder from various other nationalities. "La inmigración y emigración en México durante el último año económico," *El Economista Mexicano* 53 (October 1911–March 1912): 271.
74. Italy, Japan, the United States, Germany, Spain, and France sent special diplomatic missions. The following countries sent special envoys: Honduras, Bolivia, Austria, Cuba, Costa Rica, Russia, Portugal, Holland, Guatemala, El Salvador, Peru, Panama, Brazil, Belgium, Chile, Argentina, Norway, and Uruguay. Switzerland, Colombia, and Venezuela commissioned residents in Mexico to represent them. Great Britain could not send its mission due to the recent death of King Edward VII, nor was Nicaragua officially represented because of a recent *coup d'état*. Yet the Nicaraguan poet Rubén Darío, who had been appointed before the coup, was present and treated as a guest of honor. Mauricio Tenorio Trillo, "1910 Mexico City: Space and Nation in the City of the Centenario," *Journal of Latin American Studies* 28 (February 1996): 90; Michael J. Gonzales, "Imagining Mexico in 1910: Visions of the Patria in the Centennial Celebration in Mexico City," *Journal of Latin American Studies* 39 (2007): 511. Neither article mentions the presence of a Chinese delegation, yet some newspapers at the time referred to the Chinese special delegate for the centennial. Cortés, *Relaciones*, 116, uses the term "special ambassador" to refer to the head of the Chinese delegation. This suggests that China sent at least one special envoy to the centennial.
75. See Frederick Starr, *Mexico and the United States: A Story of Revolution, Intervention, and War* (Chicago: University of Chicago Press, 1914), 47, 54 for a description of the monuments inaugurated by the US and some European legations.
76. This practice continues today.

the surrounding buildings) to watch the fireworks and witness Porfirio Díaz's 11 p.m. speech and ringing of the independence bell from the National Palace's central balcony.[77]

From across the Pacific, only Japan and China, the sole Asian countries with diplomatic relations with Mexico, sent delegates. The former was one of six countries—together with the United States, Spain, France, Germany, and Italy—that dispatched a special mission to the centennial. Its five members included Viscount Yasuya Uchida, then ambassador to the United States and head of the delegation, who was accompanied by his wife, lieutenant colonel Kunishigue Tanaka, navy captain Tokutaro Jiraga, and Soichi Takajashi, secretary at the Mexican embassy. They disembarked from TKK's *Hong Kong Maru* on the Mexican coast and took the train to Mexico City, arriving on September 4. Lorenza R. Braniff, whose deceased husband had made a large fortune as a *hacendado*,[78] hosted them in one of the most opulent mansions on the central Paseo de la Reforma avenue. Days earlier, the Japanese legation's head, Konaichi Joriguchi, and Porfirio Díaz had inaugurated the Japanese Exhibit. Conceived during the spring visit of TKK's president Shintaro Morimoto, whose company ended up sponsoring the transportation of the objects, it comprised one of the largest displays put together for the centennial by a foreign delegation. The showcase consisted of dozens of objects that represented the vanguard of Japan's industries, arts, and agriculture. It also included a Japanese garden, a tea hall, and a small theater where concerts and martial arts performances took place. The Japanese Exhibit was hosted at the impressive Crystal Palace,[79] and it remained open to the public through the end of October.[80] The Chinese legation organized a smaller exhibit, referred to as the Salon chino (Chinese Hall), located in one of the halls of the National Palace. It included a few pieces of furniture, two fine earthenware jars, a chest, and several tapestries, and it remained open to the public after the centennial. They also donated the so-called *reloj chino*, a slim tower crowned at the top by a clock, which was placed in the Walk of Bucareli, near the Ciudadela, one of the most transited plazas of the downtown core of Mexico City. Additionally, on September 20, the Chinese ambassador hosted a special dinner at his residence for all foreign delegates and the Díaz Cabinet. Another remarkable celebration held by Chinese was that hosted by the merchant community of Mazatlan. After months of planning, they organized four days of festivities in their headquarters, the Asiatic Club, which was aptly decorated with Chinese and Mexican flags hanging from the roof. They gave away beer and lunches and, on September 18,

---

77. Gonzales, "Imagining Mexico," 496–521; Tenorio, "1910 Mexico City," 76–77.
78. A *hacendado* is a large landowner.
79. The name comes from the crystal and steel structure, built in Germany in 1895 and sent to Mexico to serve as a venue for exhibits. Once the Japanese Exhibition ended, the venue was transformed into the National Museum of Natural History. Tenorio, "1910 Mexico City," 81.
80. "Una exposición japonesa tendrá lugar en México," *El Imparcial*, March 10, 1910, 1; "Impresiones de viaje para el Diario del Pacífico," *Diario del Pacífico*, October 24, 1910, 3; Cortés, *Relaciones*, 114–16; Tenorio, "1910 Mexico City," 81, 90.

when the Sinaloa governor dropped by Mazatlan, they organized a special banquet in his honor and set off colorful fireworks ordered from San Francisco especially for the occasion. Additionally, they donated four iron columns, one for each corner of the city's central plaza, marked with the inscription "Chinese Colony of Mazatlan, 1910."[81]

Most foreign diplomats, and the foreign press, described the centennial celebrations as a great success. This was by design: Porfirian circles treated them with utmost care, making sure they visited the most catered places, circulated in the most prosperous avenues, and had limited—if any—contact with the vicissitudes of the average Mexican. The guests saw the crowds at the various public events but rarely from close up. Instead, the wealthy hosted the diplomats in the city's finest mansions. The foreign press similarly were lodged in the most expensive hotels, with all expenses paid—which amounted to 54,611 pesos of a total of 187,986 pesos that comprised the federal centennial fund.[82] The *New York Times*' correspondent wrote a glowing review: "Mexico's celebration of the 100th anniversary of martyred Father Hidalgo's proclamation of independence has been coupled with an equally impressive celebration of the eightieth anniversary of the birth of that wonderful old man, Porfirio Díaz. Who can doubt that the supposedly lesser includes the seemingly greater? Mexico's centennial of independence is unquestionably another manifestation of the power of the president."[83] Shintaro Morimoto expressed himself similarly en route to Tokyo. At a conference held at the local Chamber of Commerce he praised Díaz and dismissed the incipient rumors of possible armed dissidence. In December 1910 Díaz appointed his own son to serve as a special ambassador as an expression of gratitude to Emperor Mutsujito.[84] He never made the trip, however. In the end Morimoto was wrong. The rumors of discontent were well founded and an armed revolution against Díaz had begun.

Revolutions and the Transformation of Mexico's Transpacific Links

While the impressive centennial commemoration put together by the Porfirian government appeared to be, in the words of the *New York Times*, "another manifestation of the power of the president,"[85] there were pockets of dissent that suggested

---

81. "La colonia china y el centenario," *Diario del Pacífico*, August 25, 1910, 2; "Arte chino en México," *Diario del Pacífico*, August 31, 1910, 6; "El gobierno chino obsequia al mexicano," *Diario del Pacífico*, September 2, 1910, 1; "El obsequio de la colonia china," *Diario del Pacífico*, September 5, 1910, 1; "El Sr. Redo y el Club de Asiáticos," *Diario del Pacífico*, September 18, 1910, 1; "La colonia china obsequia al Gral. Díaz," *Diario del Pacífico*, September 21, 1910, 1; "El embajador chino da un banquete," *Diario del Pacífico*, September 21, 1910, 1; "Obsequio de China a México," *Diario del Pacífico*, November 3, 1910, 3.
82. Gonzales, "Imagining Mexico," 506. See Starr, *Mexico and the United States*, 55–57, for a description of the editorial staff who were sent to cover the centennial celebrations.
83. Gonzales, "Imagining Mexico," 521.
84. Cortés, *Relaciones*, 104, 116.
85. Gonzales, "Imagining Mexico," 521.

that Díaz was perhaps not as dominant as the affiliated press declared. Despite the protective atmosphere distancing most foreigners from the large crowds attending all public events, the expressions of discontent did not escape some visitors. This was the case, for instance, of US anthropologist Frederick Starr, who, on September 11, on his return from the inauguration of the statue of George Washington—a gift donated by US residents in Mexico—witnessed a "demonstration of 'the opposition.'" In a book published four years later, he described the event in the following terms:

> For the most part, it was a band of common working people, men and women; there were, however, a number of well-dressed men among them. Their conduct was irreproachable. From their banners we saw that they represented various anti-re-election societies . . . The groups carrying their beautiful floral pieces and a dozen or so banners began to sing the national hymn before they should march and deposit their offering in memory of the patriot fathers. [When, all of a sudden,] Castro, chief of the mounted police, face flaming and sword raised, rode into the party upon his horse, in a rage, demanding, "who is leader here?" There was no response and he ordered his men to disperse the crowd.[86]

The German ambassador in Mexico, Karl Bunz, unwittingly witnessed another anti-Díaz protest. While enjoying the September 15 fireworks preceding Díaz's speech from a balcony at the National Palace, he heard gunshots coming from the crowded plaza. They were being fired by a group of people carrying a large portrait of Francisco Madero, an opposition leader. After seeing the alarmed German, Federico Gamboa, sub-secretary of foreign relations, reassured Bunz deceitfully, lying that those were pro-Díaz demonstrators.[87]

Public criticism had increased in the electoral year of 1910 when the octogenarian Díaz announced his candidacy for president for the eighth time. The most widespread and articulate challenge came from a group led by Francisco Ignacio Madero González, a young, wealthy businessman from the northern state of Coahuila, who had studied in Mexico, France, and the United States. In his 1908 bestseller, *The Presidential Succession of 1910*, Madero assessed the state of politics in Mexico. While he praised the economic growth achieved in the previous decades, he condemned the social inequalities and the prolonged centralization of power in Díaz's hands. This, in his view, threatened the country's stability. He requested free elections without Díaz's participation. The president himself had first proposed this in a February 1908 interview with US journalist James Creelman, from *Pearson's Magazine*. Díaz would change his mind, however, announcing his nomination a few months later. Madero formed the Anti-Reelectionist Party and was elected its presidential candidate. He traveled extensively throughout the country and was widely

---

86. Starr, *Mexico and the United States*, 50–51.
87. Gonzales, "Imagining Mexico," 521–22.

welcomed throughout his campaign. In June, just days before the vote, Madero was arrested while visiting Monterrey, the country's northernmost industrialized city. He was initially accused of concealing Roque Estrada, his personal secretary, who was charged with misdemeanors. He was then taken to neighboring San Luis Potosí, where he was later accused of sedition and incarcerated for the duration of the electoral process. Díaz was proclaimed winner but accusations of fraud were widespread, particularly among Madero's supporters. After being released on bail, Madero escaped to the United States where, in early October, just as the centennial celebrations ended, he launched what became known as the Plan of San Luis Potosí, a document drafted during the time, and named after the place where he was jailed. In it he declared the elections illegal, proclaimed himself temporary president until new elections could be held, and called on the Mexican people to take up arms on November 20, at 6 p.m., to depose the current government.[88]

The armed struggle that would eventually lead to Porfirio Díaz's resignation began promptly. Armed insurrections proliferated, mostly in the northern states. While Madero's allies led some uprisings, local leaders, whose social demands went beyond the Maderista plea for free and democratic elections, also rallied. This was the case, for instance, of Emiliano Zapata, whose forces rebelled in the state of Morelos, just south of Mexico City, demanding the dissolution of large estates and the redistribution of land among the peasants. The Flores Magón brothers, anarchist leaders of the Mexican Liberal Party who operated in the borderland between the Californias, went so far as to suggest the abolition of the state and private property. As weeks passed, several rebel leaders from the north began to join forces with Madero, delivering major setbacks to Díaz's troops. In May 1911, after taking Ciudad Juárez, they forced Díaz and his vice president, Ramón Corral, to resign. The functionaries left their foreign affairs secretary as interim president with the sole purpose of organizing elections. On May 29, Díaz and his close family members traveled from Mexico City to Veracruz by train, escorted by General Victoriano Huerta. Two days later Díaz embarked for Europe aboard the German steamer *Ypiranga*, never to return.[89]

Madero won the October elections and assumed the presidency a month later;[90] however, his government was short-lived and turbulent. He had to respond

---

88. Francisco Madero, "Plan de San Luis Potosí," *Memoria Política de México*, accessed October 30, 2023, https://www.memoriapoliticademexico.org/Textos/6Revolucion/1910PSL.html.
89. The 8,103-ton *Ypiranga* belonged to the oldest German transatlantic steamship company, the Hamburg-Amerikanische Packetfahrt Aktien Gesellschaft (Hapag), which by 1911 covered the Veracruz-Havana-Vigo-Gijon-Santander-Plymouth-Le Havre route. It took Díaz twenty-four days to reach his destination. After visiting several European countries and Egypt, Díaz settled in Paris, where he died and was buried on July 2, 1915. Andrés Becerril, "Se cumplen 100 años de la partida de Porfirio Díaz en el Ypiranga," *Excelsior*, May 29, 2011, https://www.excelsior.com.mx/index.php?m=nota&id_nota=740696.
90. For an introduction to the Mexican Revolution, see Alan Knight, *The Mexican Revolution* (Lincoln: University of Nebraska Press, 1986); Javier Garciadiego and Sandra Kuntz Ficker, "La Revolución Mexicana," in *Nueva Historia General de México*, ed. Bernardo García Martínez and Javier Garciadiego (Mexico City: El Colegio de México, 2010), 537–94.

to multiple and contradictory demands that were ultimately unreconcilable. For instance, in the state of Morelos, Madero aligned with the *hacendados* who had supported his call for free elections but refused to submit to the agrarian demands of the Zapatistas. He sent federal troops to suppress the peasant revolt. This put him in direct confrontation with Zapata's forces but, unable to defeat them, he faced the *hacendados'* ire as well. The federal army was able to contain various revolts from unsatisfied ex-*maderista* allies but, in the end, several of its members ended up turning their backs on Madero. This was the case of General Huerta who, in February 1913, led a successful *coup d'état* aided by Felix Díaz—Porfirio Díaz's nephew—and Henry Lane Wilson, the Republican ambassador to Mexico, who sought more advantageous concessions for US interests. Madero, his vice president, and other close members of his political entourage were apprehended. During the fight that led to Madero's imprisonment, the recently gifted *reloj chino* was severely damaged.

The Japanese legation was then forced to play a critical role. Before his arrest, Madero had designated his brother Gustavo as special ambassador to Japan. Like Díaz's son earlier, Gustavo never made it to Japan: he was arrested and shot by Huerta on February 18. President Madero and his vice president were killed four days later. The rest of Madero's family sought refuge with the Japanese legation. Minister Kunaichi Joriguchi welcomed them. Besides impeding their arrest, he likely saved their lives.[91] The country was plunged into a decade-long civil war after Madero's assassination.

At first the circulation of steamers in the Mexican Pacific was hardly affected. Between Porfirio Díaz's May 1911 departure and Madero's February 1913 assassination, at least ten major companies continued operations in Mexico: Germany's Kosmos, with its long route between Hamburg and San Francisco; four companies from the US: PMSS, navigating weekly along the San Francisco–Panama route, Jebsen Line, which served between Seattle and Central America, Fast Steamer Salvador, renamed Salvador Railway Co. Ltd. Steamship Service, with a semi-weekly steamer between Acajutla, El Salvador, and Salina Cruz, and AHSC, navigating between Honolulu and Salina Cruz; Mexican Costa del Pacífico, circulating monthly between San Francisco and Baja Californian ports, and Compañía Naviera del Pacífico, with various national routes along the stretch between Baja California and Colima; Canadian Mexican Pacific Steamship Line (CMPS), sailing monthly between Victoria, British Columbia, and Salina Cruz with stops in Guaymas, Mazatlan, San Blas, Manzanillo, and Acapulco; and finally CCSC and TKK, each with a monthly and later bimonthly trip between Hong Kong and Salina Cruz (the latter continued all the way to Chile).[92] Some experienced delays or a

---

91. Cortés, *Relaciones*, 113–14; Rosa Elvira Vargas, "Embajador apremia a auxiliar a Japón," *La Jornada*, April 6, 2011, 23.
92. Compañía de la Guía Oficial S.A., *Guía Oficial de los Ferrocarriles y Vapores Mexicanos*, monthly issues between January 1911 and June 1913; Dirección General de Correos, *Noticia del Movimiento Probable de*

slight reduction of offerings, but circulation remained regular. The major upset was for Jebsen and TKK, whose governmental subsidies stopped abruptly. President Madero discontinued the policy in 1911 for two reasons: he needed the money to pay the federal army's increasing expenses, and his government was not deaf to a revolutionary nationalist discourse that questioned granting privileges to foreigners during Porfirian times.[93]

Nevertheless, after Madero's assassination, the civil war flared, and the Pacific coastline would feel its repercussions. A key event that triggered changes in the maritime world was the US invasion of the port of Veracruz. As mentioned above, William Taft's Republican government's support—particularly through the participation of ambassador Henry Lane Wilson—of the *coup d'état* had been key to Huerta's success. Yet within a month Democrat Woodrow Wilson assumed the presidency of the United States and changed the country's policy towards Huerta. He replaced Wilson with John Lind and demanded that the general immediately hold free elections without his own participation to restore the constitutional order and put an end to the civil war. Huerta did not accept the demand. This put him at odds with an important foreign ally and arms supplier. In April 1914, following an incident between Huerta's forces and a group of US marines in the port of Tampico, and knowing that a shipment of weapons for Huerta aboard *Ypiranga*—the same German steamer that took Díaz into exile—was about to land in Veracruz, Wilson ordered the invasion of the port, which remained under the control of US forces until November.[94] Nine days later, General Vigueras, head of Manzanillo's federal troops, received information that *Raleigh* and two other US war steamers were anchored close by. Fearing a massive landing of US marines, this time in the Pacific, he set fire to the port's pier, one of the most modern in the Mexican Pacific and the only one besides Salina Cruz that received direct transpacific shipments. In the end, the US did not invade Manzanillo. The port remained without proper docking facilities until 1952.[95]

Vapores, monthly issues between January 1911 and January 1912; "In the Business World," *The New York Times*, September 12, 1913.

[93]. René De La Pedraja, *Oil and Coffee: Latin American Merchant Shipping from the Imperial Era to the 1950s* (Westport, CT: Greenwood Press, 1998), 53–54.

[94]. On the relation between the US invasion of Veracruz and the defense of US business interests in Mexico, see John Skirius, "Railroad, Oil and Other Foreign Interests in the Mexican Revolution, 1911–1914," *Journal of Latin American Studies* 35, no. 1 (February 2003): 25–51. See also Friedrich Katz, *The Secret War in Mexico: Europe, the United States, and the Mexican Revolution* (Chicago: University of Chicago Press, 1981); Alan Knight, *U.S.–Mexican Relations, 1910–1940: An Interpretation* (La Jolla: Center for U.S.–Mexican Studies, University of California San Diego, 1987).

[95]. A rustic provisional pier was built during the 1930s. The construction of a permanent facility only started in 1946, and it was finally inaugurated in 1952. José Luis Ezquerra de la Colina, *Historia y futuro del desarrollo turístico y portuario del litoral en Manzanillo* (Estado de México: COEDI, 2006), 60. Ochoa, "Manzanillo," 123. For information on the plans of a US invasion of Manzanillo, see interview with historian Servando Ortholl in Pedro Zamora, "Según reportes de 1908 a 1914 de la Oficina de Inteligencia Naval estadounidense," *Proceso*, May, 15, 1999. In her graph, Busto, *Pacífico mexicano*, 338, suggests that the number of vessels arriving in Manzanillo did not decrease significantly after this incident, but only after 1918.

Meanwhile, the nation's second transpacific port, Salina Cruz, was also under siege. The Tehuantepec railway, which had continued to run regularly throughout the war due to the relatively low intensity of the conflict in southern Mexico, was attacked by local revolutionaries or "bandits" in April 1914.[96] Additionally, during the month following the US invasion of Veracruz, AHSC had six freighters berthed at opposite ends of the isthmus. Fearing reprisals from Huerta's government, the company rerouted all its vessels so they would no longer use Mexican ports. Instead, the regular shipments of sugar transported from Hawai'i to the east coast of the United States would now travel through the Strait of Magellan. This took twenty days longer but required no rehandling of cargo and eliminated the risk of seizure or delay by the Mexican government and local revolutionaries.[97] This left Tehuantepec without one of its most important customers. Three years later, TKK also rerouted some of its sailings out of Salina Cruz after the Japanese legation received multiple complaints of robberies.[98]

The rerouting of AHSC vessels was also motivated by the inauguration of a new and more favorable transcontinental route. After ten years of construction—or thirty-four if we consider the unsuccessful initial French efforts—by some 40,000 workers,[99] the US government finally inaugurated the Panama Canal on August 15, 1914, with the sailing of *Ancon* described at the beginning of this chapter. The Tehuantepec system, which up to then had increasingly monopolized transcontinental exchanges, suffered a major collapse. In effect, between 1913 and 1914, profits fell by more than half and, as of 1917, the system began operating at a loss.[100] In the specific case of the port of Salina Cruz, the irreversible decline in maritime movement became visible as of 1918.[101] The reasons why the Panama Canal triumphed over Tehuantepec as the preferred transcontinental route were multiple. First, the US government lobbied US steamship companies—which made up the bulk of Tehuantepec's clients—to switch.[102] This was the case, for instance, of AHSC, whose vessels only circulated via the Strait of Magellan for four months and, just as the canal opened, moved to Panama.[103] Second, the new route did not require vessels to transfer their cargo and passengers. Third, its rates were more

---

96. De la Pedraja, *Oil and Coffee*, 55. "Bandits" was a polemical term used by various groups to discredit those who opposed them or whom they viewed as being against their interests.
97. The first regular run via the Strait of Magellan began on April 29. Thomas C. Cochran and Ray Ginger, "The American–Hawaiian Steamship Company, 1899–1919," *The Business History Review* 28, no. 4 (December 1954): 360.
98. AHGE-SRE, 16-20-176, "Robos de equipajes de japoneses que han desembarcado en Salina Cruz," July 1917, 1, 9.
99. John W. Herbert, "The Panama Canal: Its Construction and Its Effect on Commerce," *Bulletin of the American Geographical Society* 45, no. 4 (1913): 250.
100. Vera Valdés Lakowsky, *De las minas al mar. Historia de la plata mexicana en Asia, 1565–1834* (Mexico City: FCE, 1987), 21; Arce, *Transportes*, 146–48.
101. See graph in Busto, *Pacífico mexicano*, 314.
102. Arce, *Transportes*, 148.
103. They once again moved to the Strait of Magellan Strait when the canal was temporarily closed in 1915–1916 due to landslides. Cochran and Ginger, "American–Hawaiian," 361.

competitive: while Tehuantepec transhipment costs averaged about $3.50 per ton of cargo, the Panamanian toll required less than a third of that amount.[104] Finally, as mentioned above, the Tehuantepec isthmus presented additional uncertainties related to the Mexican Revolution.

World War I, which also began in August, also had an impact on maritime traffic. In the Pacific, most commercial carriers had to either stop circulating, reduce their sailings, or transform their load, purpose, and schedules to serve the war effort. Kosmos represents an example of the former: the line linking Hamburg with South American, Central American, and Mexican Pacific ports since the turn of the century stopped as the Germans withdrew from Pacific trade.[105] AHSC exemplifies the latter: the company chartered some of its vessels to other companies for the war as early as 1914. The same was done by the Panama Railroad Company, the owner of *Ancon*, which also ended up transporting war-related cargo. By 1916 AHSC had suspended its intercoastal transport and its ships were operating only as war charters. Additionally, half of its fleet had switched to cover transatlantic runs. A year later—once the US formally entered the war—and until 1919, its entire fleet passed under government control. Throughout this time, AHSC vessels carried a million tons of cargo to the Allies, and at the end of the war, they returned over 122,000 US soldiers from Europe. Even after the war ended, the firm never reopened its Hawaiian route.[106]

PMSS, the company that had first started a regular transpacific passenger service in 1867, also underwent change. Following passage of the Seamen's Act in 1915, which required all US maritime companies to ensure that 75 percent of their crew members were able to understand English commands, PMSS's general manager, Rennie P. Schwerin, declared that this would bankrupt his company. The new law was made to get rid of the Chinese crews who had traditionally been employed by US maritime companies—particularly those servicing Asia. PMSS announced the sale of its fleet and set its final sailing for August. In the end, it was saved by capital injected by the American International Corporation and W. R. Grace & Co. Relief, however, was only temporary. PMSS was ultimately unable to compete with transpacific companies with newer and stronger fleets and, following the loss of a bid for five larger vessels in 1925, it ceased operations and was acquired by Dollar Co., the company that won the same bid.[107]

The wars also affected the two direct regular passenger offerings between Mexico and Asia, which were first reduced and then, in 1916, canceled. CCSC, the company that had inaugurated regular transpacific trips to Mexico in 1903, had been engaged in a dispute with the Mexican government since 1908. As explained

---

104. G. G. Huebner, "Economic Aspects of the Panama Canal," *The American Economic Review* 5, no. 4 (December 1915): 817.
105. Huebner, "Economic Aspects," 817.
106. Cochran and Ginger, "American–Hawaiian," 361–63.
107. Tate, *Transpacific Steam*, 39–41, 78.

in the previous chapter, it demanded compensation for the more than 900 steerage passengers who were not allowed to disembark after Salina Cruz's health delegate had declared them infected with a contagious eye disease—a diagnosis that the company contested.[108] While Díaz had informally expressed his desire to settle the claim in a friendly manner, Madero's government explicitly refused to provide compensation. In 1913, as both parties faced contract renewal and with the confrontation at its apex—and the country in chaos—it seems highly unlikely that this took place.[109] The disappearance of all reference to CCSC from the Official Guide for Mexican Railways and Steamers as of June 1913 seems to confirm this assumption.[110] Yet there is evidence that CCSC continued with irregular sailings for at least another year, but with a much reduced list of steerage passengers.[111] This was also the case for the only other regular transpacific steamship company existing at the time. TKK had been affected by the actions of the revolutionary government since 1911, when Madero canceled subsidies to foreign maritime companies, which, in its case, amounted to $10,000 per round trip.[112] It had continued sailing, albeit with a bimonthly— rather than monthly—itinerary.[113] During that year the authorities registered 4,910 Chinese entries to Mexico. But in 1914 the number dropped to 1,491.[114] By then both TKK and CCSC vessels were crossing the ocean with fewer and fewer passengers. This was the case, for instance, of TKK's *Anyo Maru*, which in August 1914 brought only sixty travelers—two in first class, seven in second class, and fifty-one in steerage.[115] Other examples include CCSC's *Marie* and *Mexico City*, which docked in Manzanillo in July and August 1914 with 245 and 37 steerage passengers respectively.[116] In 1915 and 1916, Chinese arrivals diminished again, numbering 474 and 228 respectively. In the case of the Japanese, only 337 entries were registered during the 1911–1920 period.[117] Steamship passenger service from Asia to Mexico shut down in 1916 due to the world war. When business reopened after the war in 1919, 1,151 Chinese entered the country. The number increased to 2,669 in 1920 and dropped to 1,320 in 1921. Whereas Chinese immigration experienced

---

108. See Chapter 5 for more details.
109. The original 1903 contract had a five-year duration, extendable indefinitely every five years until either of the signing parties objected. "Contrato," *Diario Oficial*, February 1903, 677.
110. Compañía de la Guía Oficial S.A., *Guía Oficial de los Ferrocarriles y Vapores Mexicanos* 13, no. 7 (June 1913).
111. No records were found of CCSC arrivals to the Mexican coast after 1914.
112. AHGE-SRE, L-E-187, 55.
113. For 1913, the company listed three vessels—*Kiyo Maru*, *Anyo Maru*, and *Buyo Maru*—circulating between Asian ports, Manzanillo, and Salina Cruz in March, May, July, September, November, and January. Compañía de la Guía Oficial S.A., *Guía Oficial de los Ferrocarriles y Vapores Mexicanos* 13, no. 3 (February 1913): 83.
114. Chao, *Chinese in Mexico*, 54.
115. Archivo General de la Nación de México (AGNM), Galería 5, Ramo Gobernación, Periodo revolucionario, caja 123, expediente 44, "El Sr. Lic. Luis Riba, comunica que el vapor 'Anyo Marú' de la línea Toyo Kisen Kaisha llegará al país procedente del Japón, con sesenta pasajeros," August 3, 1914.
116. *Marie*, once again like *Suisang* in 1908 (see Chapter 5), had to spend several weeks docked at port waiting for the health inspector's clearance. See AGNM, Galería 5, Ramo Gobernación, Periodo revolucionario, caja 123, expedientes 43, 45, 46, June–August 1914.
117. Ota, "Características sociales," 60.

a resurgence once transpacific steamship trade recommenced, it never rebounded to prewar levels. Whereas 12,114 Chinese entered Mexico between 1911 and 1913, only 10,062 did so between 1919 and 1928, and the numbers continued to drop.[118] As for the Japanese, records show 1,636 entries in the 1920s. However, in 1926, TKK, unable to face the increasing competition of US and Canadian transpacific companies, merged with the strongest, government-backed Japanese enterprise, Nippon Yusen Kaisha (NYK).[119] While it is not certain what happened with the Mexico/South America itinerary once NYK took over, the truth is that Japanese immigration to Mexico would never again reach prewar levels.

The dramatic decrease of Asian steerage passengers arriving in Mexico—the central business of transpacific steamship ventures up to then—was related not only to the emergence of the world war but also to the accentuation of a nationalist economic policy and of a xenophobic discourse on the part of Mexican revolutionary governments.[120] These trends were already visible by the end of the Porfiriato as the government began to shift its policies,[121] enacting its first ever restrictive Immigration Law right after CCSC's 1908 claim described in Chapter 5. While a new immigration authority was created in 1909 to enforce this law and keep better track of foreigners, ultimately it was unable to carry out its duties due to the revolutionary chaos.[122] In the year of Díaz's exile, José María Romero, one of the members of the 1903 commission,[123] published a 121-page report on Chinese. He concluded that it was not "advisable for the national interests to permit the unlimited and unrestricted immigration of Chinese as an element of colonization [. . . nor] as an element of manual labor, be it in group or individual form, free or by contracts formed outside of our territory."[124] He discredited the idea of Chinese as industrious laborers and rather dismissed them as an unproductive "race," notably inferior to the more desirable, hardworking Europeans. Additionally, he argued that they remained loyal to Asia and therefore were not prone to assimilation. Even worse, if they integrated into Mexican society through their daily coexistence with the local lower classes, they would only degenerate the masses with their unhygienic practices and delay the much-desired assimilation of Indigenous communities to the idealized citizen, that is, a Spanish-speaking and Europeanized Mestizo. Following this trend, the federal government enacted a new Immigration Law in 1926 that regulated foreign presence more thoroughly. It gave the secretary of the interior

---

118. Chao, *Chinese in Mexico*, 54–55.
119. Tate, *Transpacific Steam*, 68.
120. Elliott Young, *Alien Nation: Chinese Migration in the Americas from the Coolie Era through World War II* (Chapel Hill: University of North Carolina Press, 2014), 197.
121. On this subject, see Paul Garner, *British Lions and Mexican Eagles: Business, Politics, and Empire in the Career of Weetman Pearson in Mexico, 1889–1919* (Stanford, CA: Stanford University Press, 2011), 94–137.
122. Young, *Alien Nation*, 201; Grace Peña Delgado, *Making the Chinese Mexican: Global Migration, Localism, and Exclusion in the U.S.-Mexico Borderlands* (Stanford, CA: Stanford University Press, 2012), 102.
123. The 1903 commission to study the pertinence of Chinese arrivals in Mexico is described in Chapter 5.
124. Cited in Chao, *Chinese in Mexico*, 181.

special powers to establish temporary prohibitions on the entry of foreign workers and granted greater powers to sanitation agents to deny entry to immigrants than the Porfirian 1908 law did.[125] By this time, Japanese laborers had practically ceased traveling to Mexico. Instead, it was mostly family members of those already living there and young professionals—notably doctors, dentists, veterinarians, and pharmacists—who arrived in the country, after a 1917 bilateral treaty encouraged this type of migration.[126] Chinese, meanwhile, continued to land—although in decreasing numbers—aboard TKK vessels or via the irregular sailings of companies such as Spain and China Navigation Company, owned by merchants from Hong Kong and operated mostly by Spanish and Philippine sailors profiting from the transportation of Cantonese to Spanish-speaking countries, notably Peru, Chile, and Cuba.[127] In the early 1930s, in the midst of an economic crisis, Mexican legislation further intensified its xenophobic, protectionist trend, enacting laws that temporarily barred the entrance of immigrant laborers and required businesses to employ 90 percent Mexican nationals. In addition, a new report on Asian immigration blamed Chinese once again for introducing health hazards and stealing locals' jobs, thereby forcing foreign laborers to migrate to the United States. All these reports and legislative acts contributed to putting an end to the arrival of Asian workers in Mexico by the early 1930s.[128]

Laborers were not the only ones affected by the revolutionary anti-foreign discourses and policies that equated the "evil" Porfiriato with the privileged treatment of foreigners—to the detriment of Mexicans. So too were merchants, particularly those of Chinese descent.[129] While they had faced discrimination during the Porfiriato, it was not until the beginning of the Revolution that they became the targets of systematic attacks. As early as 1911, Chinese and Japanese merchants from across the country filed complaints with their embassies—which, in turn, sent urgent petitions to the secretary of foreign affairs to request protection for their subjects—after their businesses were ransacked by revolutionary troops or uncontrolled mobs.[130] The most horrendous episode—and the worst act of violence experienced by any Chinese diasporic community in the Americas during the twentieth century—happened in the northern city of Torreon. In May 1911, after Maderista forces took hold of the city, they killed 303 Chinese and 5 Japanese and caused close to $850,000 of damage to their businesses. As of 1916, a more organized anti-Chinese movement began to consolidate with the creation of leagues and juntas, particularly in the northeastern states. Formed by middle-class Mexican merchants and often supported by poor local workers, they sought to eliminate the competition

---

125. Chao, *Chinese in Mexico*, 44–45.
126. Ota, "Características sociales," 57.
127. Chao, *Chinese in Mexico*, 44–45.
128. Chao, *Chinese in Mexico*, 180–90.
129. By 1926 Chinese had become the second largest resident foreign ethnic community in the country, with 24,218 registered members after Spaniards, who amounted to 48,558. Chao, *Chinese in Mexico*, 55.
130. See, for instance, AHGE-SRE, 16-4-33; 16-4-34; 16-4-56; 16-5-115; 16-4-60; 13-1-143; 13-1-145; 13-1-149.

of Chinese merchants and their employees. They appealed to patriotic propaganda that portrayed Chinese as foreigners who had become rich at the expense of Mexican small entrepreneurs and workers. The former could not compete with the allegedly corrupt business practices of the Chinese, and the latter found themselves displaced from jobs by foreigners who worked for paltry salaries. As had become standard practice, Chinese were also blamed for introducing diseases and vices, as well as degrading the "Mexican race" when procreating with local women. The hostility was exacerbated by the economic crisis in the late 1920s, when Chinese served as the scapegoat for worsening economic conditions. Tension reached its peak in the summer of 1931, when the Sonoran governor issued an order to expel Chinese merchants due to their non-compliance with a state labor law requiring that 80 percent of employees be Mexican-born. The Chinese fought back through legal means but, in October, a new order forbade them from reopening their businesses and accused them of violating an existing state law that barred marriage between Chinese men and Mexican women. As a result, thousands of Chinese merchants and their Mexican families left Sonora for good. While a few relocated to other states, the large majority left the country.[131] The Chinese population never recovered from this blow and, by 1940, only 4,859 Chinese lived in Mexico.[132]

One final factor that contributed to altering the configuration of the Pacific maritime world was the introduction of oil technology. Just as with steam, the Anglo empires—first Britain, then the United States—began implementing oil technology in their maritime industries. While oil exploration can be traced to the 1860s, it was not until the first decade of the twentieth century that the systematic study and exploration of oil deposits began in earnest. The British led the way in converting naval and merchant vessels from coal to oil. By 1911, when Winston Churchill became the First Lord of the Admiralty and assumed command of the Royal Navy, the country already had over sixty destroyers and seventy-four submarines running on oil. Most of its vessels had been equipped with oil sprayers in lieu of coal furnaces. The US government followed closely and, in 1913, the newly appointed Wilson administration ordered the construction of its first oil-fueled battleship. By the end of that year, the US Navy had built or had under construction four battleships, thirty submarines, and forty-one destroyers, all of which were oil-fueled. It was simultaneously progressively switching the remaining fleet to oil. The

---

131. Chao, *Chinese in Mexico*, 145–90. See also Juan Puig Llano, *Entre el río Perla y el Nazas: La China decimonónica y sus braceros emigrantes, la colonia china de Torreón y la matanza de 1911* (Mexico City: Conaculta, 1992). See Julia María Schiavone Camacho, *Chinese Mexicans: Transpacific Migration and the Search for a Homeland, 1910–1960* (Chapel Hill: University of North Carolina Press, 2012) to learn about Chinese Mexicans in the decades following those covered by this book.
132. Roberto Ham Chande, "La migración china hacia México a través del Registro Nacional de Extranjeros," in *Destino México: Un estudio de las migraciones asiáticas a México, siglos XIX y XX*, ed. María Elena Ota Mishima (Mexico City: El Colegio de México, 1997), 179. See the extensive literature on many of these themes cited by Jian Gao, "Restoring the Chinese Voice during Mexican Sinophobia, 1919–1934," *The Latin Americanist* 63, no. 1 (March 2019): 48–72.

change was particularly useful in the Pacific, where coal supplies were limited and oil reserves were abundant.[133] That is why, in regard to commercial vessels, it was a Pacific firm, AHSC, that was at the vanguard of oil technologies. In 1902, it had installed an oil burner—the so-called Lassoe-Lokevin, in honor of its inventors—in two of its newest steamers, *Nevada* and *Nebraska*, running from San Francisco to Hawai'i. That summer the former completed the first ocean voyage by an oil-burning vessel under the US flag. Two years later, *Nebraska* became the first oil-powered vessel to engage in intercoastal trade with its pioneering fifty-two-day trip between San Diego and New York via the Strait of Magellan. The *Marine Journal* referred to this accomplishment as "the sensation of the time" and listed the following advantages: "taking into consideration the saving of time, cargo space, and fire room force, it has been estimated that this vessel saved on the voyage about $20,000 in the substitution of oil for fuel in place of coal."[134] Other benefits included more efficient combustion,[135] longer and faster trips without the need to refuel, and the possibility of greater cargo volume as oil storage needs were much reduced compared with coal. On the Mexican coastline, the revolution indirectly enhanced the switch from coal to oil as imports of coal diminished and became more expensive, while oil began to be more available as local deposits were exploited by Mexican and UK firms. Therefore, by the mid-1910s, many local and foreign ship owners sent their vessels to the US for the relatively inexpensive procedure of swapping out coal-fueled boilers for those running on oil.[136]

The gradual conversion to oil had various repercussions for the maritime traffic on the Mexican Pacific coastline. To begin with, vessels running along the San Francisco–Panama route no longer needed to stop in Mexican ports to refuel. Second, ports that possessed adequate oil facilities were preferred over those that did not, leading to the emergence of new ports and the decay of others. Third, the advent of oil as fuel promoted the creation of new communication technologies, such as cars and planes, which in some cases replaced, and in others competed with, more traditional shipping routes. Fourth, oil technologies promoted the construction of infrastructure that affected ports unevenly. This was the case of roads and highways, which eventually replaced the routes covered by railroads in Mexico. Consequently, ports with highway connections might have acquired more pre-eminence. Acapulco, for instance, became the first Pacific port to be connected to Mexico City via a paved road, in 1927.[137] Lastly, new geopolitics in the Pacific—and the world—developed in the competition for control of oil supplies, of which Mexico possessed substantial reserves.

---

133. John A. DeNovo, "Petroleum and the United States Navy before World War I," *The Mississippi Valley Historical Review* 41, no. 4 (March 1955): 641–56.
134. Cited in Cochran and Ginger, "American–Hawaiian," 349.
135. As opposed to coal, with oil, the boiler doors were not opened throughout the trip and therefore the heat remained constant, which enhanced the boiler's longevity and efficiency.
136. De la Pedraja, *Oil and Coffee*, 54.
137. Busto, *Pacífico mexicano*, 296, 403–4.

Conclusions: Mexico and the Transpacific World in the 1910s

In 1910, when Porfirio Díaz celebrated his eightieth birthday, his eighth re-election, and the country's centennial, Mexico's Pacific coast had more regular maritime connections than ever before. Over a dozen steamship companies linked its ports internationally, particularly with San Francisco and Panama. Additionally, two monthly direct transpacific runs by CCSC and TKK had prospered, particularly due to the transportation of Asians in steerage to Mexico. The two companies served the route between Hong Kong, Yokohama, Manzanillo, and Salina Cruz—with additional stopovers when and if there was enough demand. The latter two had consolidated as the only Mexican ports with regular transpacific connections. This was due to their modern infrastructure and railroad connections. The former connected to two of the country's largest cities, Mexico City and Guadalajara, and the latter boasted a short and fast transcontinental connection via the Tehuantepec railroad. While the country still lacked shipyards capable of building transpacific vessels and various national liners that would have made it less dependent on the variable nature of foreign actors, the situation seemed full of promise, much like the impressive centennial celebrations put together by the Porfirian regime, attended by some thirty foreign delegations, including those of China and Japan, and hundreds of thousands of local and international spectators.

But the apparent macroeconomic progress was not founded on solid social or political grounds. When different groups revolted and demanded more rights in response to Francisco Madero's call to arms, the regime fell within months. The Mexican Revolution erupted and a period of instability that had repercussions for Mexico's transpacific links followed. To begin with, both Manzanillo and Salina Cruz and their connecting railroads suffered; their infrastructure and reputation deteriorated due to armed attacks combined with chaotic administration. While this did not immediately arrest maritime traffic, its decline came gradually and by 1918 was irreversible. Additionally, the two transpacific steamship companies offering direct transport to Mexico went from reducing to suspending to finally disbanding operations by the 1920s.

The Mexican Revolution also contributed to the decline in maritime traffic along its Pacific coast and to the eventual extinction of the main transpacific business during the Porfiriato, that is, the transportation of Asians in steerage. This was due to the exacerbation of two tendencies present at the end of the period: a more nationalist economic policy combined with xenophobic, anti-Chinese discourse, actions, and legislation. The revolutionary governments scapegoated the Chinese for the economic crisis. Chinese, who had increased their presence and participation in the economy during the Porfiriato, particularly in the north, were accused of taking wealth and jobs from Mexicans as well as polluting the locals' health. During the harshest period of armed confrontation, these immigrants suffered both personal and property attacks. Once the armed violence diminished, Chinese businessmen became the main targets of a nationalist and xenophobic legislation that obliged them to fire Chinese workers and hire Mexicans. The state used the same legislation to fine the Chinese for alleged public health violations, forbid them from marrying local women, seize their properties, and finally, in 1931, expel them from Sonora, the state where most of them lived. The local Chinese population would never recover from this blow.

The Japanese presence also dwindled. Nationalist revolutionary governments stopped subsidizing foreign shipping companies. This affected TKK's economic interests as their initial contract stipulated an important income from the Mexican government. Moreover, their main business was no longer viable due to the xenophobic events and laws described above. Additionally, direct transport was interrupted due to hostilities associated with the World War I and, in the 1920s, the company ended up disappearing. In addition, after the 1910s the Japanese no longer migrated in large groups, but rather as young professionals, sometimes with family, following a 1917 bilateral agreement.

The August 1914 opening of the Panama Canal and the outbreak of the world war would irrevocably affect Mexico's transpacific connections. In the case of the former, most of the transcontinental traffic that had formerly used Salina Cruz and the Tehuantepec railroad shifted operations. The reasons were plentiful: the canal offered a short, inexpensive, and peaceful gateway to the Atlantic, and it was heavily lobbied for by the US government, particularly for its own merchant navy. In addition, the logistic challenges occasioned by the outbreak of the war reduced the merchant fleet circulating in the Pacific as many vessels, among them *Ancon*, were channeled from their original commercial routes to operate as war charters, resulting in the cancelation of regular transpacific passenger service between Mexico and Asia from 1916 to 1919.

A technological revolution delivered the final blow to the Porfirian Pacific maritime system based mostly on the circulation of coal-fired steamers. Beginning in the 1910s, oil, slowly but surely, took over as the main maritime fuel. *Ancon*, for instance, was refitted from coal to oil-firing between 1919 and 1920.[138] The appearance of oil introduced new geopolitics, routes, ports, needs, infrastructure, and communication technologies that would, just as had happened during the transition from sail to steam, once again transform the face of Mexican transpacific connections, a topic that is beyond the scope of the present work.

---

138. "Ancon I," The Ship Stamp Society, last modified October 15, 2008, https://shipstamps.co.uk/forum/viewtopic.php?t=7169.

Conclusion

In the age of the steamship, people and objects moved like never before. In turn, they set off a chain of reactions that washed over the world like the wake left behind the steamers themselves. These waves touched everything, crossing localities, borders, oceans, and continents, changing ideas and minds in spite of—and in some cases because of—the status quo. This book situates itself in the midst of these tides and has used Elizabeth Sinn's proposal of "the in-between place" as a paradigm for migration studies instead of the more traditional notions of places of origin, destinations, or their interconnections in order to showcase steamships' key role in promoting movement and transforming lives across the Pacific as of the second half of the nineteenth century.

By concentrating on the steamships that circulated "in-between" Asian and American coasts, this book exposed how these new technologies reconfigured power dynamics and accentuated mobility by compressing time and space. The first were those circulating in the mid-nineteenth century, overloaded with male passengers lured by the promises of riches emanating from Pacific coast gold rushes. Ships such as *Monumental City*, *New Orleans*, and *Golden Age* turned an arduous, months-long journey by sail into a straight-line voyage that could transport thousands of people and mountains of cargo in just a handful of weeks. Fueled by coal and human ambition, these motors of change were a consequence of the consolidation of industrial capitalism dovetailing with Anglo-imperial expansionism. The changes wrought together by these traveling people and vessels consisted not only in surmounting previously technical and geographical obstacles but also in supercharging a nascent industrial economy, including the construction of railroads, ports, factories, and the development of specialized machinery, and in introducing an upward spiral of wants and needs.

Steamships embodied the values of modernity and helped mold people's vision of themselves and of others as individuals and collectives, contributing to the formation of personal, national, regional, and imperial imaginings, aspirations, prejudices, and inequalities. This process was accentuated by the stratified treatments and procedures to which the ships' diverse passengers were subjected. These primarily depended on class (first, second, or steerage) and were exacerbated once "race," economic status, and nationality were factored in, as was evident on the *Suisang*, where the Chinese in steerage were forced to undergo a series of compulsory examinations where ultimately many were refused landing and thus indirectly sentenced to die on the return voyage, while sick second-class passengers disembarked freely. The pride and astonishment of Chapter 1's first-class passenger finding herself moving quickly against the wind past a sea of smaller sailing vessels offers another example, capturing some of this sense of individual entitlement, but also of neglect for those with lesser means. Steamers thus affected the dreams and routines of everyone they passed, even of the large crowds that regularly gathered at the port for welcomes and send-offs. While these latter experiences were not the subject of this book, they hint at the impact that vessels had on the surrounding populations. Steamers also manifested national and imperial prestige in displays of geopolitical power. Perry's 1853–1854 expeditions aboard *Mississippi*, *Susquehanna*, and *Powhatan* forced Japan to sign a series of unequal treaties. In turn, Japan's 1910 pan-Pacific voyage of *Asawa* and *Kasagi* paraded the country's expansionist policy in Asia. *Ancon*'s 1914 sailing of the Panama Canal, a concession originally given for life to the US government by a nascent Panamanian state, offered a statement beyond the obvious symbol of inauguration.

Individuals, interest groups, communities, and emerging nations throughout the Pacific Rim participated in the reconfiguration of the transpacific mosaic, each stamping their imprint with varying levels of access and success, depending on their position in the rising and ebbing cartographies of power. Certainly, the British and US governments, situated at the cusp of industrial and technological advancements, consolidated their geopolitical influence in the Pacific. The former appropriated its strategically located Hong Kong outpost in the 1830s, while the latter defeated the Mexican army in 1848, taking over much of its extensive west coast at the timely moment of the discovery of gold in California. Both empires also had access to capital surpluses and the foresight to sponsor and defend costly commercial steamer fleets. The British subsidized the Pacific Steam Navigation Company (PSN), operating out of South America as of 1840. Eight years later, the US-supported Pacific Mail Steamship Company (PMSS) inaugurated its Panama to Oregon route. In 1867 it then connected California with Yokohama and Hong Kong, buoying San Francisco into importance as North America's largest international transport hub.

Yet the Pacific was far from a "British" or "American Lake," as some within a more imperial tradition have previously sustained. The diverse interests of Mexican, Chinese, and Japanese diplomats, sailors, passengers, bureaucrats, businessmen, merchants, and laborers helped form the contours of transpacific life. For instance, the establishment of the two companies that thwarted PMSS's transoceanic monopoly, the Hong Kongese China Commercial Steamship Company (CCSC) and the Japanese Toyo Kisen Kaisha (TKK), complexified transpacific maritime networks by directly connecting Mexico and Asia at the beginning of the twentieth century. The routes departing from Manzanillo and Salina Cruz helped transform the latter into a key international port, particularly in the fruitful period between the opening of the transcontinental Tehuantepec railroad and the inauguration of the Panama Canal. While Mexican attempts to create a transpacific line failed, the country's systematic participation in the network dates from as early as 1848, when it provided ports of call. When PMSS launched its transoceanic line, Mexico's ports connected with Asia via San Francisco. Furthermore, as the Mexican Astronomic Commission's trip showed, they also connected via the Atlantic through Veracruz, using the Gulf of Mexico steamers and then the US transcontinental railroad to San Francisco.

Notwithstanding these clear instances, global histories, Pacific studies, and maritime and national historiographies have rarely acknowledged Mexico's participation and links with Asia during the age of steam. Instead, they have for the most part sustained that their interactions ended in 1815 with the sailing of the last Manila galleon, in the context of the War of Independence from Spain. In conjunction with pioneering research, mostly from Asian American, borderlands, diaspora, and transnational studies, this book has insisted on highlighting the longevity and importance of Mexican–Asian connections and the participation of Mexican, Chinese, and Japanese interests in the construction of a Pacific region.

As the examples in the book have shown, decisions by hegemonic actors that affected Pacific networks were neither overpowering nor monolithic, but rather characterized by shifting alliances and antagonisms. These depended not only on imperial, national, or ethnic affiliations, but also on class, gender, age, profession, economic interests, and notoriously on "race," an elastic notion used by those in power to justify their privileges within a hierarchical social order and to discriminate against those considered inferior. For instance, the stakeholders behind the Compañía Mexicana de Navegación del Pacífico (CMNP) forged alliances with London bureaucrats to circumnavigate the sailing prohibition imposed by Hong Kong's imperial authorities. Later, the company's British representatives confronted their own compatriots who had chartered *Mount Lebanon* to the CMNP. Mexican authorities, on their part, supported their co-nationals' pleas at times but disregarded them at others. The CCSC's Hongkongese stakeholders, who worked together with Mexican authorities, suffered the same fate once the latter denied entry to most of *Suisang*'s steerage passengers. The 1908 dispute also saw the country's doctors facing off against one another and eventually drew in Chinese and British diplomats as well. Long-lasting bonds were created between Mexican and Japanese authorities who were sympathetic to each other's modernizing projects and demands for equal treatment on the international stage. These consonances led to unprecedented bilateral relations, which in turn paved the way for the arrival of the first Japanese state-sponsored colonists to the Americas aboard *Gaelic*. In contrast, many Mexican elites ended up embracing racist ideologies against Chinese, as was visible in the Astronomic Commission's visit to Asia aboard *Vasco de Gama* in 1874. Initially the Chinese were allowed into the country only as cheap labor for large industrializing projects, such as the building of the Salina Cruz port and the Tehuantepec transcontinental railroad. The authorities later used medical terminology to limit Chinese entry aboard CCSC vessels at the very port that the Chinese themselves had helped build and, eventually, in the 1930s, expelled them from some northern states.

In spite of the regulations and restrictions imposed by imperial and national elites, common travelers found ways around them and fueled the formation of transpacific networks. As we have seen, the main business sustaining transpacific steamship service from the 1860s to the 1910s was transporting Cantonese in steerage. Ostensibly they were being sent to work as underpaid laborers, but thousands arrived in Mexico with a very different goal: to reach the United States through its unsupervised border, particularly after the 1882 Exclusion Act. Many who remained in Mexico sidestepped the Sinophobe traps laid in their paths, becoming prosperous merchants. These, in turn, exchanged consumables with others to improve their lot, supporting, yet again, the circulation of steamers. Likewise, the state-sanctioned Japanese colonists abandoned imperial plans soon after arrival, creating their own prosperous businesses and associations that subsequently brought more Japanese to Latin America. In terms of the ships themselves, regardless of the place of origin of the owners and operators, Chinese, Japanese, and Mexican sailors formed the main workforce aboard transpacific carriers. The movement and presence of all these peoples in this vast sphere broke norms, materialized changes, and made it possible for these ships to cross oceans. Without these actors, the Pacific as we know it would not exist.

In sum, by tracing the routes and stories around a selected group of transpacific steamers between 1867 and 1914, this study has shown how they became sites of contention where shifting power relations among Cantonese, Japanese, Mexican, British, and US individuals and collectives clashed, converged, and were constantly (re)negotiated. In order to complexify these stories, it is important to continue adding layers of historical experiences from other peoples, steamers, and commodities, keeping in mind the generative value of the journeys in-between.

Bibliography

Archives and Libraries

Archivo General de Indias (AGI), Seville
Archivo General de la Nación de México, Mexico City (AGNM)
Archivo General del Estado de Oaxaca (AGEO)
Archivo Histórico de Mazatlán, Mazatlan (AHM)
Archivo Histórico de la Secretaría de la Secretaría de Comunicaciones y Transportes, Mexico City
Archivo Histórico de la Secretaría de Marina Armada de México, Mexico City
Archivo Histórico del Estado de Colima, Colima City
Archivo Histórico del Municipio de Colima, Colima City
Archivo Histórico del Municipio de Manzanillo, Manzanillo, Colima
Archivo Histórico Genaro Estrada, Secretaría de Relaciones Exteriores, Mexico City (AHGE-SRE)
Archivo Personal de Horacio Archundia, Manzanillo, Colima
Archivo Personal de Javier Juárez Yamamoto, Acacoyagua, Chiapas
Archivo Personal Matías Romero (APMR), Instituto Mora, Mexico City
Biblioteca Central de la Universidad Nacional Autónoma de México (UNAM), Mexico City
Biblioteca Daniel Cosío Villegas, El Colegio de México, Mexico City
Biblioteca José María Lafragua, SRE, Mexico City
Biblioteca Miguel Lerdo de Tejada (Fondo Reservado, Fondo Histórico de Hacienda & Hemeroteca), Mexico City
Hemeroteca Nacional de México (Fondo Reservado), UNAM, Mexico City
J. Porter Shaw Library, San Francisco
Mapoteca Manuel Orozco y Berra, Mexico City
National Archives and Record Administration (NARA), San Bruno, California
Public Records Office, Hong Kong (PRO-HK)
San Francisco Maritime Museum Library and Archives, San Francisco
Sutro Library, San Francisco
The National Archives, London
University of British Columbia Library

Primary Sources

Maps

Calderón, Francisco A. *Carta postal y de vías de comunicación de los Estados Unidos Mexicanos [Mexican Postal Chart and Communication Routes]*. Map. Mexico City: Dirección General de Correos, 1910. From Mapoteca "Manuel Orozco y Berra," República Mexicana 7. https://mapoteca.siap.gob.mx/cgf-rm-m27-v7-0333/ (accessed December 15, 2023).

Gorsuch, Robert B. *Líneas de ferrocarriles en los Estados Unidos del Norte y México [Railway Lines in the United States and Mexico]*. Map. Mexico City: V. de Murguía e hijos, 1871. From Mapoteca "Manuel Orozco y Berra," Internacionales 2. https://mapoteca.siap.gob.mx/coyb-int-m50-v2-0055/ (accessed December 15, 2023).

Jaimes, M. *Plano del lote que pertenece a la colonia Japonesa en el Soconusco [Map of the Land Belonging to the Japanese Colony in Soconusco]*. Map. Chiapas: 20th century. From Mapoteca "Manuel Orozco y Berra," Chiapas 6. https://mapoteca.siap.gob.mx/cgf-chis-m2-v6-0499/ (accessed December 15, 2023).

Jiménez, Alfredo A. y Carlos Vega Schiafino. *Carta general de vías y comunicaciones de los Estados Unidos Mexicanos [General Map of Mexican Roads and Communications]*. Map. Mexico City: Secretaría de Comunicaciones, 1907. From Mapoteca "Manuel Orozco y Berra," República Mexicana 5. https://mapoteca.siap.gob.mx/cgf-rm-m26-v5-0285/ (accessed December 15, 2023).

Merino, Vicente E. *Ruta de las flotas de España para América y la Nao de América para China [Route followed by the Spanish Fleets to America and from America's Nao to China]*. Map. Mexico City: H. Iriarte, 1879. From Mapoteca "Manuel Orozco y Berra," Internacionales 3. https://mapoteca.siap.gob.mx/coyb-int-m50-v3-0075/ (accessed December 15, 2023).

Vías de comunicación marítimas y terrestres [Maritime and Land Routes]. Map. Mexico City, E. Andriveau Coujon, 1882. From Mapoteca "Manuel Orozco y Berra," Planisferios 1. https://mapoteca.siap.gob.mx/cgf-planf-m35-v1-0064/ (accessed December 15, 2023).

Periodicals

China Mail, 1884–1885
Daily Bee, 1867
Diario del Hogar, 1897
Diario del Pacífico, 1910–1911
Diario Oficial, 1903
El Correo de Chihuahua, 1903–1910
El Economista Mexicano, 1886–1915
El Estado de Colima, 1906
El Imparcial, 1897–1910
El Monitor Republicano, 1891–1896
El País, 1899–1900
El Tiempo. Diario Católico, 1883–1910
Guía Oficial de los Ferrocarriles y Vapores Mexicanos, 1902–1913

Hong Kong Daily Press, 1902, 1904, 1907
La Gaceta Marítima, 1896
Movimiento Probable de Vapores, 1903–1912
New York Daily Herald, 1865
New York Times, 1864, 1896, 1900–1905

Books and Pamphlets

Biblioteca Virtual Miguel de Cervantes. "México a través de los siglos." Last modified 2017. https://www.cervantesvirtual.com/nd/ark:/59851/bmctj0m8. (Original volumes from 1882).

Bulnes, Francisco. *El porvenir de las naciones latinoamericanas ante las recientes conquistas de Europa y Norteamérica. Estructura y evolución de un continente.* Mexico City: El pensamiento vivo de América, 1899.

Bulnes, Francisco. *Sobre el Hemisferio Norte, once mil leguas: impresiones de viaje a Cuba, los Estados Unidos, el Japón, China, Conchinchina, Egipto y Europa.* Mexico City: Revista Universal, 1875.

Commercial Relations of the United States with Foreign Countries during the Years 1885 and 1886, Vol. 1. Washington: Government Printing Office, 1887.

De Garay, José. *An Account of the isthmus of Tehuantepec in the Republic of Mexico; with Proposals for Establishing a Communication between the Atlantic and Pacific Oceans, Based upon the Surveys and Reports of a Scientific Commission, Appointed by the Projector, Don José de Garay.* London: J.D. Smith & Co., 1848.

Díaz Covarrubias, Francisco. *Viaje de la Comisión Astronómica Mexicana al Japón para observar el tránsito del planeta Venus por el disco del Sol el 8 de diciembre de 1874.* Mexico City: Políglota, 1876.

Further Correspondence respecting the Renewal of Diplomatic Relations with Mexico. London: Foreign Office, 1885.

Gemelli Careri, Giovanni Francesco. *A Voyage to the Philippines.* Manila: Filipiniana Book Guild, 1963. (Original from 1700).

International Chinese Business Directory of the World for the Year 1913. San Francisco, CA: International Chinese Business Directory Co., 1913.

Lloyd's Register of British and Foreign Shipping. London: Wyman and Sons, 1881.

Madero, Francisco. "Plan de San Luis Potosí." Memoria Política de México. Accessed October 30, 2023. https://www.memoriapoliticademexico.org/Textos/6Revolucion/1910PSL.html.

Reglamento y Aranceles Reales para el Comercio Libre de España a Indias de 12 de Octubre de 1778. Madrid: 1778.

Sierra, Justo. *Evolución política del pueblo mexicano.* Caracas: Biblioteca Ayacucho, 1985. (Original texts from 1900–1902).

Toyo Kisen Kaisha/Línea Oriental de Vapores. *Folleto sobre el servicio directo entre Japón, China, México y Sudamérica.* 1909.

Secondary Sources

Akachi, Jesús K., Carlos T. Kasuga, Manuel S. Murakami, María Elena Ota Mishima, Enrique Shibayama, and René Tanaka. "Japanese Mexican Bibliographic Essay." In *Encyclopedia of Japanese Descendants in the Americas: An Illustrated History*, edited by Akemi Kikumura-Yano, 222. Lanham, MD: Rowman & Littlefield, 2002.

Alonso, Ana María. "Conforming Disconformity: 'Mestizaje,' Hybridity, and the Aesthetics of Mexican Nationalism." *Cultural Anthropology* 19, no. 4 (November 2004): 459–90.

Andrien, Kenneth J., and Lyman L. Johnson. *The Political Economy of Spanish America in the Age of Revolution, 1750–1850*. Albuquerque: University of New Mexico Press, 1994.

Arce Ibarra, Roxana. *Los transportes en el istmo de Tehuantepec*. Mexico City: UNAM, 1949.

Avilés Galán, Miguel Ángel. "A todo vapor: Mechanization in Porfirian Mexico. Steam Power and Machine Building, 1862–1906." PhD diss., University of British Columbia, 2010.

Avilés Galán, Miguel Ángel. "Measuring Skulls: Race and Science in Vicente Riva Palacio's México a través de los siglos." *Bulletin of Latin American Research* 29, no. 1 (January 2010): 85–102.

Azuma, Eiichiro. *Between Two Empires: Race, History, and Transnationalism in Japanese America*. New York: Oxford University Press, 2005.

Azuma, Eiichiro. "Historical Overview of Japanese Emigration, 1868–2000." In *Encyclopedia of Japanese Descendants in the Americas: An Illustrated History*, edited by Akemi Kikumura-Yano, 32–33. Lanham, MD: Rowman & Littlefield, 2002.

Basave Benítez, Agustín F. *México mestizo: análisis del nacionalismo mexicano en torno a la mestizofilia de Andrés Molina Enríquez*. Mexico City: FCE, 1992.

Beaglehole, J. C. *The Exploration of the Pacific*. London: A. & C. Black, 1934.

Bean, Walton, and James J. Rawls. *California: An Interpretive History*. New York: McGraw Hill, 1988.

Beddie, M. K. *Bibliography of Captain James Cook*. Sydney: The Library of New South Wales, 1970.

Becerril, Andrés. "Se cumplen 100 años de la partida de Porfirio Díaz en el Ypiranga." *Excelsior*, May 29, 2011, https://www.excelsior.com.mx/index.php?m=nota&id_nota=740696.

Behdad, Ali. "The Politics of Adventure: Theories of Travel, Discourses of Power." In *Travel Writing, Form, and Empire: The Poetics and Politics of Mobility*, edited by Julia Kuehn and Paul Smethurst, 80—94. New York: Routledge, 2009.

Bellwood, P. S. "The Peopling of the Pacific." *Scientific American* 243, no. 5 (1980): 1–17.

Benítez, Fernando. *El galeón del Pacífico: Acapulco–Manila, 1565–1815*. Chilpancingo: Gobierno del Estado de Guerrero, 1992.

Bentley, Jerry H. "Sea and Ocean Basins as Frameworks of Historical Analysis." *Geographical Review* 89, no. 2 (1999): 215–24.

Bernabeu Albert, Salvador. *El Pacífico Ilustrado: del Lago Español a las grandes expediciones*. Madrid: MAPFRE, 1992.

Bernstein, Harry. *Matías Romero, 1837–1898*. Mexico City: FCE, 1973.

Bernstein, Marvin D. *The Mexican Mining Industry, 1890–1950*. New York: State University of New York Press, 1964.

Bjork, Katharine. "The Link That Kept the Philippines Spanish: Mexican Merchant Interests and the Manila Trade, 1571–1815." *Journal of World History* 9, no. 1 (1998): 25–50.

Bonialian, Mariano A. *La América española: entre el Pacífico y el Atlántico: globalización mercantil y economía política, 1580–1840*. Mexico City: El Colegio de México, 2019.

Bonilla, Juan de Dios. *Historia marítima de México*. Mexico City: Litorales, 1962.

Brookes, Barbara, Warwick Anderson, and Miranda Johnson, eds. *Pacific Futures: Past and Present*. Honolulu: University of Hawai'i Press, 2018.

Buffington, Robert M., and William E. French. "The Culture of Modernity." In *The History of Mexico*, edited by Michael C. Meyer and William H Beezley, 398–400. New York: Oxford University Press, 2000.

Busto Ibarra, Karina. "El espacio del Pacífico mexicano: puertos, rutas, navegación y redes comerciales, 1848–1927." PhD diss., El Colegio de México, 2008.

Busto Ibarra, Karina. *El Pacífico mexicano y sus transformaciones: integración marítima y terrestre en la configuración de un espacio internacional, 1848–1927*. Mexico City: El Colegio de México, 2022.

Butler, Shannon Marie. *Travel Narratives in Dialogue: Contesting Representations of Nineteenth-Century Peru*. New York: Peter Lang, 2008.

Campbell, Persia Crawford. *Chinese Coolie Emigration to Countries within the British Empire*. London: BiblioLife, 2009.

Cárdenas de la Peña, Enrique. *Historia de las Comunicaciones y Transportes en México: Marina mercante*. Mexico City: Secretaría de Comunicaciones y Transportes, 1988.

Carrillo Martín, Rubén. "Los 'chinos' de Nueva España: Migración asiática en el México colonial." *Millars Espai i Història* 21, no. 39 (January 2015): 15–40. https://core.ac.uk/download/pdf/132357334.pdf.

Carroll, John M. *Edge of Empires: Chinese Elites and British Colonials in Hong Kong*. Cambridge, MA: Harvard University Press, 2005.

Carter, George F. "Movement of People and Ideas across the Pacific." In *Plants and the Migrations of Pacific Peoples: A Symposium*, edited by Jacques Barrau, 7–22. Honolulu: Bishop Museum Press, 1961.

Cathcart, A. H. "Pacific Mail—Under the American Flag Around the World." *Pacific Marine Review* (July 1920): 53–58.

Chao Romero, Robert. *The Chinese in Mexico, 1882–1940*. Tucson: University of Arizona Press, 2010.

Chaves, José Ricardo. "Estudio preliminar: Bulnes viajero." In *Francisco Bulnes. Sobre el hemisferio norte, once mil leguas. Impresiones de viaje a Cuba, los Estados Unidos, el Japón, China, Conchinchina, Egipto y Europa*, edited by Coordinación de Humanidades, 7–26. Mexico City: UNAM, 2012.

Chávez Jiménez, Daniar. "Viajeros del siglo XIX: el linaje mexicano y las 11 mil leguas de Francisco Bulnes por el hemisferio norte." *Estudios* 108, no. 12 (Spring 2014): 53–72.

Chere, Lewis M. *The Diplomacy of the Sino-French War (1883–1885): Global Complications of an Undeclared War*. Notre Dame: Cross Cultural Publications, 1988.

Chong, José Luis. "Hijo de un país poderoso. La inmigración china a América (1850–1950)." *Diacronías. Revista de Divulgación Histórica* 1, no. 1 (February 2008): 55–64.

Coatsworth, John. "Obstacles to Economic Growth in Nineteenth-Century Mexico." *American Historical Review* 83 (1978): 80–100.

Cochran, Thomas C., and Ray Ginger. "The American-Hawaiian Steamship Company, 1899–1919." *The Business History Review* 28 (December 1954): 343–65.

Conner, Glen. *History of Weather Observations: San Francisco, California, 1844–1948*. Ashville, NC: Midwestern Regional Climate Center, 2005. https://mrcc.purdue.edu/files/FORTS/histories/CA_San_Francisco_Conner.pdf.

Connolly, Priscilla. *El contratista de Don Porfirio. Obras públicas, deuda y desarrollo desigual*. Mexico: ColMich/UAM/FCE, 1997.

Cortés, Enrique. *Relaciones entre México y Japón durante el Porfiriato*. Mexico City: SRE, 1980.

Cott, Kenneth. "Mexican Diplomacy and the Chinese Issue, 1876–1910." *The Hispanic American Historical Review* 67, no. 1 (February 1987): 63–85.

Craib, Raymond B. III. *Chinese Immigrants in Porfirian Mexico: A Preliminary Study of Settlement, Economic Activity and Anti-Chinese Sentiment*. Latin American Institute, University of New Mexico, Research Paper Series 28. Albuquerque: University of New Mexico, 1996.

Crump, Thomas. *The Age of Steam: The Power that Drove the Industrial Revolution*. London: Constable & Robinson, 2007.

Cumberland, Charles C. "The Sonora Chinese and the Mexican Revolution." *The Hispanic American Historical Review* 40, no. 2 (May 1960): 191–211.

Cushner, Nicholas P. *Spain in the Philippines: From Conquest to Revolution*. Manila: Ateneo de Manila University, 1971.

Dambourges Jacques, Leo M. "The Chinese Massacre in Torreon (Coahuila) in 1911." *Arizona and the West* 16 (Autumn 1974): 233–46.

De La Pedraja, René. *Oil and Coffee: Latin American Merchant Shipping from the Imperial Era to the 1950s*. Westport, CT: Greenwood Press, 1998.

De la Torre Villar, Ernesto. *Lecturas Históricas Mexicanas*, Vol. 2. Mexico City: IIH-UNAM, 1998. https://historicas.unam.mx/publicaciones/publicadigital/libros/lecturas/T2/LHMT2_067.pdf.

De Miguel Bosch, José Ramón. "Andrés de Urdaneta y el tornaviaje." Accessed November 21, 2023. http://www.andresurdaneta.org/urdaneta500/de/biografia.asp?cod=1754&nombre=1754&nodo=&orden=True&sesion=1.

Dennis, Philip A. "The AntiChinese Campaigns in Sonora, Mexico." *Ethnohistory* 26, no. 1 (Winter 1979): 65–80.

DeNovo, John A. "Petroleum and the United States Navy before World War I." *The Mississippi Valley Historical Review* 41, no. 4 (March 1955): 641–56.

De Vega, Mercedes, Francisco Javier Haro, José Luis León, and Juan José Ramírez. *Historia de las relaciones internacionales de México, 1821–2010*, Vol. 6: *Asia*. Mexico City: SRE, 2011.

Diego, Hugo. *Viaje al Japón. Francisco Díaz Covarrubias*. Mexico City: Ediciones de Educación y Cultura, 2008.

Dong, Jingsheng. "Chinese Emigration to Mexico and the Sino-Mexico Relations before 1910." *Estudios Internacionales (Chile)* 38 (January–March 2006): 75–88.

Donmenge, Francois, Alain Hentz de Lenps, and Odile Chapnis. *Contribution Française a la connaisance géographique des Mers du Sud: Bibliographie des Principaux Travaux Scientifique Francais Traitant des Oceand Pacifique et Indien, des Mers Australes et de leurs ∂Iles*. Talence: Ceqet/Cret, 1990.

Duarte Espinosa, María de Jesús. *Frontera y diplomacia: las relaciones México-Estados Unido durante el Porfiriato*. Mexico City: SRE, 2001.
Dudden, Arthur Power. *The American Pacific: From the Old China Trade to the Present*. Oxford: Oxford University Press, 1992.
Duncan, Robert H. "The Chinese and the Economic Development of Northern Baja California, 1889–1929." *The Hispanic American Historical Review* 74, no. 4 (November 1994): 615–47.
Duncan, Roland E. "William Wheelright and Early Steam Navigation in the Pacific 1820–1840." *The Americas* 32, no. 2 (October 1975): 257–81.
Dym, Jordana, and Christophe Belaubre, eds. *Politics, Economy, and Society in Bourbon Central America, 1759–1821*. Boulder: University Press of Colorado, 2007.
Eastman, L. *Throne and Mandarins: China's Search for a Policy during the Sino-French Controversy*. Stanford, CA: Stanford University Press, 1984.
Elleman, B. *Modern Chinese Warfare, 1795–1989*. London: Routledge, 2001.
Escobar Bautista, María del Pilar. "México–Alemania: datos de una valiosa relación histórica." *Revista mexicana de política exterior* 99 (September–December 2013): 175–83.
Esposito, Matthew D. *Funerals, Festivals, and Cultural Politics in Porfirian Mexico*. Albuquerque: University of New Mexico Press, 2010.
Ezquerra de la Colina, José Luis. *Historia y futuro del desarrollo turístico y portuario del litoral en Manzanillo*. Estado de México: COEDI, 2006.
Falcón, Romana, and Raymond Buve, eds. *Don Porfirio presidente nunca omnipotente: hallazgos, reflexiones y debates*. Mexico City: Universidad Iberoamericana, 1998.
Feifer, George. *Breaking Open Japan: Commodore Perry, Lord Abe, and American Imperialism in 1853*. New York: Smithsonian/HarperCollins, 2006.
Finney, Ben. "The Other One-Third of the Globe." *Journal of World History* 5, no. 2 (1994): 273–97.
Flynn, Dennis, and Arturo Giráldez. "Cycles of Silver: Globalization as Historical Process." *World Economics* 3, no. 2 (April–June 2002): 1–16.
Flynn, Dennis, and Arturo Giráldez. "Los orígenes de la globalización en el siglo XVI." In *Oro y plata en los inicios de la economía global: de las minas a la moneda*, edited by Bernd Hausberger and Antonio Ibarra, 29–76. Mexico City: El Colegio de México, 2014.
Flynn, Dennis O., Arturo Giraldez, and James Sobredo, eds. *Studies in Pacific History: Economics, Politics, and Migration*. Burlington, VT: Ashgate, 2002.
Fradera, Josep M., and Luis Alonso, eds. *Imperios y naciones en el Pacífico*. Madrid: Asociación Española de Estudios del Pacífico/Consejo Superior de Investigaciones Científicas, 2001.
French, William E. "Imagining and the Cultural History of Nineteenth-Century Mexico." *The Hispanic American Historical Review* 79, no. 2 (May 1999): 249–67.
French, William E. *A Peaceful and Working People: Manners, Morals, and Class Formation in Northern Mexico*. Albuquerque: University of New Mexico Press, 1996.
Friel, Ian. *Maritime History of Britain and Ireland c. 400–2001*. London: British Museum Press, 2003.
Frost, Alan, and Jane Samson, eds. *Pacific Empires: Essays in Honour of Glyndwr Williams*. Melbourne: Ashgate, 1999.

Frost, Lionel. "Rim of Fire: Pacific Rim Cities and the Problem of Fires." In *Studies in Pacific History: Economics, Politics, and Migration*, edited by Dennis O. Flynn, Arturo Giráldez, and James Sobredo, 108–22. Burlington, VT: Ashgate, 2002.
Gallegos López, Sergio Armando. "Los fletes marítimos: factor de desarrollo en el comercio exterior de México." BA thesis, Universidad Nacional Autónoma de México/FCPyS, 1972.
Gao, Jian. "Chinese Migration to Latin America: From Colonial to Contemporary Era." *History Compass* 19, no. 9 (2021): 1–13.
Gao, Jian. "Restoring the Chinese Voice during Mexican Sinophobia, 1919–1934." *The Latin Americanist* 63, no. 1 (March 2019): 48–72.
García Benavides, Roberto. *Hitos de las comunicaciones y los transportes en la historia de México*. Mexico City: Secretaría de Comunicaciones y Transportes, 1988.
García Triana, Mauro, and Pedro Eng Herrera. *The Chinese in Cuba, 1847–Now*. Lanham, MD: Lexington Books, 2009.
Garciadiego, Javier. *Textos de la Revolución Mexicana*. Caracas: Fundación Biblioteca Ayacucho, 2010.
Garciadiego, Javier, and Sandra Kuntz Ficker. "La Revolución Mexicana." In *Nueva Historia General de México*, edited by Bernardo García Martínez and Javier Garciadiego, 537–94. Mexico City: El Colegio de México, 2010.
Garner, Paul. *British Lions and Mexican Eagles: Business, Politics, and Empire in the Career of Weetman Pearson in Mexico, 1889–1919*. Stanford, CA: Stanford University Press, 2011.
Garner, Paul. *Porfirio Díaz*. London: Pearson, 2001.
Garner, Paul. *Porfirio Díaz: Entre el mito y la historia*. Mexico City: Crítica, 2015.
Garner, Richard L. *Economic Growth and Change in Bourbon Mexico*. Gainesville: University Press of Florida, 1993.
Glade, William. "Latin America and the International Economy, 1870–1914." In *The Cambridge History of Latin America*, Vol. 4, edited by Leslie Bethell, 1–56. Cambridge: Cambridge University Press, 1989.
Glick Schiller, Nina, Linda Basch, and Cristina Szanton Blanc. "From Immigrant to Transmigrant: Theorizing Transnational Migration." *Anthropological Quarterly* 68, no. 1 (January 1995): 48–63.
Gómez Izquierdo, José Jorge. *El movimiento antichino en México (1871–1934): problemas del racismo y del nacionalismo durante la Revolución Mexicana*. Mexico City: INAH, 1991.
Gonzales, Michael J. "Imagining Mexico in 1910: Visions of the Patria in the Centennial Celebration in Mexico City." *Journal of Latin American Studies* 39 (2007): 495–533.
González, Fredy. *Paisanos chinos: Transpacific Politics among Chinese Immigrants in Mexico*. Oakland: University of California Press, 2017.
González Navarro, Moisés. *La colonización en México, 1877–1910*. Mexico City: Talleres de Impresión de Estampillas y Valores, 1960.
González Navarro, Moisés. *Los extranjeros en México y los mexicanos en el extranjero, 1821–1970*, Vols. 1, 2. Mexico City: El Colegio de México, 1993–1994.
González Navarro, Moisés. "Mestizaje in Mexico during the National Period." In *Race and Class in Latin America*, edited by Magnus Mörner, 145–69. New York: Columbia University Press, 1970.

Gordon, Peter, and Juan José Morales. *La plata y el Pacífico: China, Hispanoamérica y el nacimiento de la globalización, 1565–1815*. Madrid: Siruela, 2022.

Gotkowitz, Laura, ed. *Histories of Race and Racism: The Andes and Mesoamerica from Colonial Times to the Present*. Durham, NC: Duke University Press, 2011.

Government of British Columbia. "Federal Head Tax." Accessed October 28, 2023. https://www2.gov.bc.ca/gov/content/governments/multiculturalism-anti-racism/chinese-legacy-bc/history/discrimination/federal-head-tax.

Gracida Romo, Juan José. "Guaymas, notas para la historia comercial del puerto, 1820–1910." In *Los puertos noroccidentales de México*, edited by Jaime Olveda and Juan Carlos Reyes, 199–212. Zapopan: El Colegio de Jalisco, 1994.

Graham, Richard, Thomas E. Skidmore, Aline Helg, and Alan Knight. *The Idea of Race in Latin America, 1870–1940*. Austin: University of Texas Press, 1990.

Grunstein, Arturo. *Railroads and Sovereignty: Policy-Making in Porfirian Mexico*. Los Angeles: University of California Press, 1994.

Guardia, Sara Beatriz, ed. *Viajeras entre dos mundos*. Dourados: UFGD, 2012.

Gutiérrez Grageda, Blanca Estela, and Héctor P. Ochoa Rodríguez. *Las caras del poder: conflicto y sociedad en Colima, 1893–1950*. Colima: Universidad de Colima/Gobierno del Estado de Colima/Conaculta, 1995.

Gutiérrez Hernández, Adriana. "Juárez, las relaciones diplomáticas con España y los españoles en México." *Estudios de historia moderna y contemporánea de México* 34 (July–December 2007): 29–63.

Hale, Charles. *The Transformation of Liberalism in Late Nineteenth-Century Mexico*. Princeton, NJ: Princeton University Press, 1989.

Ham Chande, Roberto. "La migración china hacia México a través del Registro Nacional de Extranjeros." In *Destino México: Un estudio de las migraciones asiáticas a México, siglos XIX y XX*, edited by María Elena Ota Mishima, 167—88. Mexico City: El Colegio de México, 1997.

Hamnett, Brian. *Juárez*. New York: Longman, 1994.

Hardee, Jim. "Soft Gold: Animal Skins and the Early Economy of California." In *Studies in Pacific History: Economics, Politics, and Migration*, edited by Dennis O. Flynn, Arturo Giraldez, and James Sobredo, 23–39. Burlington, VT: Ashgate, 2002.

Harley, J. B. *The New Nature of Maps: Essays in the History of Cartography*. London: Johns Hopkins University Press, 2001.

Herbert, John W. "The Panama Canal: Its Construction and Its Effect on Commerce." *Bulletin of the American Geographical Society* 45, no. 4 (1913): 241–54.

Herrera, Inés. "Comercio y comerciantes de la costa del Pacífico mexicano a mediados del siglo XIX." *Historias* 20 (April–September 1988): 129–35.

Hoare, J. E. *Japan's Treaty Ports and Foreign Settlements: The Uninvited Guests, 1858–1899*. Kent, CT: Japan Library, 1994.

Hoerder, Dirk. *Cultures in Contact: World Migrations in the Second Millennium*. Durham, NC: Duke University Press, 2002.

Hu-DeHart, Evelyn. "Ceremonia Solemne de Recepción como Académica Corresponsal en Estados Unidos: Evelyn Hu-DeHart: Petición de perdón en Torreón y memoria histórica de los chinos en México." Filmed January 2023 at Academia Mexicana de la Historia, Mexico City. Video, 1 hour: 35 min. https://www.youtube.com/watch?v=Qk46Uk0fkas.

Hu-DeHart, Evelyn. "The Chinese in Northern Mexico, 1875–1932." *The Journal of Arizona History* 21, no. 3 (Autumn 1980): 275–312.
Hu-DeHart, Evelyn. "Latin America in Asia-Pacific Perspective." In *Asian Diasporas: New Formations, New Conceptions*, edited by Rhacel Parreñas and Lok Siu, 29–62. Stanford, CA: Stanford University Press, 2007.
Hu-DeHart, Evelyn. "Latin America in Asia-Pacific Perspective." In *What Is In a Rim? Critical Perspectives on the Pacific Region Idea*, edited by Arif Dirlik, 251–82. Boulder, CO: Westview Press, 1993.
Hu-DeHart, Evelyn. "Mexico." In *The Encyclopedia of the Chinese Overseas*, edited by Lynn Pann, 256–58. Cambridge, MA: Harvard University Press, 1998.
Hu-DeHart, Evelyn. "Multiculturalism in Latin American Studies: Locating the 'Asian' Immigrant; or, Where Are the Chinos and Turcos?" *Latin American Research Review* 44, no. 2 (2009): 235–42.
Hu-DeHart, Evelyn. "On Coolies and Shopkeepers: The Chinese as *Huagong* (Laborers) and *Huashang* (Merchants) in Latin America/Caribbean." In *Displacements and Diasporas: Asians in the Americas*, edited by Wanni W. Anderson and Robert G. Lee, 78–111. New Brunswick, NJ: Rutgers University Press, 2005.
Hu-DeHart, Evelyn. "Racism and Anti-Chinese Persecution in Mexico." *Amerasia Journal* 9, no. 2 (1982): 1–28.
Huebner, G. G. "Economic Aspects of the Panama Canal." *The American Economic Review* 5, no. 4 (December 1915): 816–29.
Hyde, Francis E. *Far Eastern Trade, 1860–1914*. London: Adam and Charles Black, 1973.
Ichioka, Yuji. *The Issei: The World of the First Generation Japanese Immigrants, 1885–1924*. New York: The Free Press, 1988.
Irie, Kiyoshi. "Nuevos relatos sobre México." *Estudios de Asia y África* 92, no. 3 (September–December 1993): 421–48.
James, Margarita. "My Transpacific Life." *BC Studies* 204 (Winter 2019–2020): 139–50.
Jansen, Marius B. *The Making of Modern Japan*. Cambridge, MA: Harvard University Press, 2000.
Jennings, J. N., and G. J. R. Linge, eds. *Of Time and Place: Essays in Honour of O.H.K. Spate*. Canberra: Australian National University Press, 1980.
Jett, Stephen. *Crossing Ancient Oceans: Voyages to the Americas before Columbus*. New York: Copernicus, 2007.
Johnson, Donald D. *The United States in the Pacific: Private Interests and Public Policies, 1784–1799*. Westport, CT: Praeger, 1995.
Katz, Friedrich. *The Secret War in Mexico: Europe, the United States, and the Mexican Revolution*. Chicago: University of Chicago Press, 1981.
Kellogg, Louise P. "The United States and Japan." *The Wisconsin Magazine of History* 4, no. 3 (March 1921): 347–49.
Kemble, John Haskell. "The Big Four at Sea: The History of the Occidental and Oriental Steamship Company." *Huntington Library Quarterly* 3, no. 3 (April 1940): 339–57.
Kemble, John Haskell. "The Genesis of the Pacific Mail Steamship Company." *California Historical Society Quarterly* 13, no. 4 (December 1934): 386–406.
Kemble, John Haskell. "The Gold Rush by Panama, 1848–1851." *The Pacific Historical Review* 18, no. 1 (February 1949): 45–56.

Kemble, John Haskell. "Pacific Mail Service between Panama and San Francisco, 1849–1851." *Pacific Historical Review* 2, no. 4 (December 1933): 405–17.
Kilingray, David, Margarette Lincoln, and Nigel Rigby, eds. *Maritime Empires: British Imperial Maritime Trade in the Nineteenth Century*. Woodbridge: Boydell Press/National Maritime Museum, 2004.
Knight, Alan. *The Mexican Revolution*. Lincoln: University of Nebraska Press, 1986.
Knight, Alan. "Racism, Revolution, and *Indigenismo*: Mexico, 1910–1940." In *The Idea of Race in Latin America, 1870–1940*, edited by Richard Graham, 71–113. Austin: University of Texas Press, 1990.
Knight, Alan. *U.S.-Mexican Relations, 1910–1940: An Interpretation*. La Jolla: Center for U.S.-Mexican Studies, University of California, San Diego, 1987.
Kobayashi, Audrey, and Midge Ayukawa. "A Brief History of Japanese Canadians." In *Encyclopedia of Japanese Descendants in the Americas: An Illustrated History*, edited by Akemi Kikumura-Yano, 150–61. Lanham, MD: Rowman & Littlefield Publishers, 2002.
Kooiman, William. "Grace's Pacific Mail, 1915–1925." *Journal of the Puget Sound Maritime Historical Society* 21, no. 1 (September 1987): 3–20.
Kuehn, Julia, and Paul Smethurst, eds. *Travel Writing, Form, and Empire: The Poetics and Politics of Mobility*. New York: Routledge, 2009.
Kuntz Ficker, Sandra. *El comercio exterior de México en la era del capitalismo liberal, 1870–1929*. Mexico City: El Colegio de México, 2007.
Kuntz Ficker, Sandra. "Fuentes para el estudio de los ferrocarriles durante el Porfiriato." *América Latina en la historia económica. Boletín de fuentes* 13–14 (January–December 2000): 137–48.
Kuntz Ficker, Sandra. "Los ferrocarriles y la formación del espacio económico en México, 1880–1910." In *Antologías de historia económica de México: Ferrocarriles y obras públicas*, edited by Sandra Kuntz and Priscilla Connolly, 105–37. Mexico City: Instituto Mora/ColMich/Colegio de México/IIH-UNAM, 1999.
Lajous, Roberta. *La política exterior del porfiriato*. Mexico City: El Colegio de México, 2010.
Laorden Jiménez, Luis. *Navegantes españoles en el Océano Pacífico*. Madrid: Taograf, 2013.
Latour, Bruno. *Reassembling the Social: An Introduction to Actor-Network Theory*. Oxford: Oxford University Press, 2005.
Lavalle Argudín, Mario. *La Armada en el México independiente*. Mexico City: Instituto Nacional de Estudios Históricos de la Revolución Mexicana/Secretaría de Marina, 1985.
Lavalle Argudín, Mario. *Memorias de Marina. Buques de la Armada de México. Acontecimientos notables, 1821–1991*, Vol. 2. Mexico City: Secretaría de Marina, 1992.
Lawson, Will. *Pacific Steamers*. Glasgow: Brown, Son & Ferguson, 1927.
Lee, Erika. *The Making of Asian America: A History*. New York: Simon & Schuster, 2015.
Lesser, Jeffrey. *A Discontented Diaspora: Japanese Brazilians and the Meanings of Ethnic Militancy, 1960–1980*. Durham, NC: Duke University Press, 2007.
Lesser, Jeffrey, ed. *Searching for Home Abroad: Japanese Brazilians and Transnationalism*. Durham, NC: Duke University Press, 2003.
Look Lai, Walton. "The Caribbean." In *The Encyclopedia of the Chinese Overseas*, edited by Lynn Pann, 248–53. Cambridge, MA: Harvard University Press, 1998.
Loriaux, Florence, Philippe Marechal, Jan Possemiers, Eddy Stols, Patricia Van Schuylenbergh-Marchand, Jean-Luc Vellut, and Luc Vints. *Les Belges et le Mexique: dix contributions*

à l'histoire des relations Belgique-Mexique. Louvain: Presses Universitaires de Louvain, 1993.
Lukacs, John. *The Future of History*. New Haven, CT: Yale University Press, 2011.
MacGregor, Josefina. *Matías Romero. Textos escogidos*. Mexico City: CNCA, 1992.
MacLachlan, Colin M., and William H. Beezley. *Mexico's Crucial Century, 1810–1910: An Introduction*. Lincoln: University of Nebraska Press, 2010.
Macleod, Julia H., Matthew Perry, George Henry Preble, and James Rodgers Goldsborough. "Three Letters Relating to the Perry Expedition to Japan." *The Huntington Library Quarterly* 6, no. 2 (February 1943): 228–37.
Mandujano López, Ruth. "Cantoneses en Manzanillo: la importancia del 'lugar de en medio' en el proceso migratorio." In *Tierra receptora y espacios de apropiación. Extranjeros en la historia de México, siglos XIX y XX*, edited by Martín López and Marcela Martínez, 321–36. Zamora: El Colegio de Michoacán/El Colegio de San Luis, 2015.
Mandujano López, Ruth. "From Sail to Steam: Coastal Mexico and the Reconfiguration of the Pacific in the 19th Century." *International Journal of Maritime History* 22, no. 2 (December 2010): 247–75.
Mandujano López, Ruth. "La migración interminable, cantoneses en Manzanillo." *Legajos. Boletín del Archivo General de la Nación* 1 (July–September 2009): 44–58.
Mandujano López, Ruth. "La redefinición del espacio transpacífico durante la era de los vapores: el caso de las compañías marítimas entre México y Asia (1884–1910)." In *Intercambios, actores, enfoques. Pasajes de la historia latinoamericana en una perspectiva global*, edited by Aarón Grageda Bustamante, 117–32. Sonora: Universidad de Sonora, 2014.
Mandujano López, Ruth. "Transpacific Mexico: Encounters with China and Japan in the Age of Steam (1867–1914)." PhD diss., University of British Columbia, 2012.
Mandujano López, Ruth. "Una revolución al vapor: la navegación transpacífica entre México y Asia a finales del siglo XIX." *Retos Internacionales* 3 (Fall 2010): 9–19.
Marder, Arthur J. "From Jimmu Tennō to Perry: Sea Power in Early Japanese History." *The American Historical Review* 51, no. 1 (October 1945): 1–34.
Marine & Oceans. "The Panama Canal Turns 109: Interesting Facts about This Strategic Location." Last modified September 8, 2023. https://marine-oceans.com/en/the-panama-canal-turns-109-interesting-facts-about-this-strategic-location/.
Marsden, Ben, and Crosbie Smith. *Engineering Empires: A Cultural History of Technology in Nineteenth-Century Britain*. New York: Palgrave Macmillan, 2005.
Martínez Legorreta, Omar. "De la modernización a la guerra." In *Japón: su tierra y su historia*, edited by Daniel Toledo, Michiko Tanaka, Omar Martínez Legorreta, Jorge Alberto Lozoya, and Víctor Kerber, 173–241. Mexico City: El Colegio de México, 1991.
Martínez Montiel, Luz María. *Inmigración y diversidad cultural en México*. Mexico City: UNAM, 2005.
Martínez Peña, Luis Antonio. "Mazatlán, historia de su vocación comercial, 1823–1910." In *Los puertos noroccidentales de México*, edited by Jaime Olveda and Juan Carlos Reyes, 157–78. Zapopan: El Colegio de Jalisco, 1994.
Martínez Shaw, Carlos. *El Pacífico Español de Magallanes a Malaspina*. Madrid: Ministerio de Asuntos Exteriores, 1988.

Masterson, Daniel, with Sayaka Funada-Classen. *The Japanese in Latin America*. Champaign: University of Illinois Press, 2004.
"Matías Romero (1837–1898)." In *Instituto Matías Romero. XXV Aniversario*, 107–40. Mexico City: SRE, 1999.
McCullough, David. *The Path between the Seas: The Creation of the Panama Canal, 1870–1914*. New York: Simon & Schuster, 1977.
McGuinness, Aims. *Path of Empire: Latin American Transformations and the California Gold Rush, 1848–1856*. Ithaca, NY: Cornell University Press, 2008.
McKeown, Adam. *Chinese Migrant Networks and Cultural Change: Peru, Chicago, Hawaii, 1900–1936*. Chicago: University of Chicago Press, 2001.
McKeown, Adam. "Global Migration, 1846–1940." *Journal of World History* 15, no. 2 (2004): 155–89.
McMaster, John. "Aventuras asiáticas del peso mexicano." *Historia Mexicana* 8, no. 3 (January–March 1959): 372–99.
McNeil, William F. *Visitors to Ancient America: The Evidence for European and Asian Presence in America Prior to Columbus*. Jefferson, NC: McFarland & Company, 2005.
Mejia, Javier. "The Economics of the Manila Galleon." *Journal of Chinese Economic and Foreign Trade Studies* 15, no. 1 (2021): 35–62.
Méndez Rodenas, Adriana. *Transatlantic Travels in Nineteenth-Century Latin America: European Women Pilgrims*. Lanham, MD: Bucknell University Press/Rowman & Littlefield, 2014.
Minger, Ralph Eldin. "Panama, the Canal Zone, and Titular Sovereignty." *The Western Political Quarterly* 14, no. 2 (June 1961): 544–54.
Minna Stern, Alexandra. "From Mestizophilia to Biotypology: Racialization and Science in Mexico, 1920–1960." In *Race and Nation in Modern Latin America*, edited by Nancy P. Appelbaum, Anne S. Macpherson, and Karin Alejandra Rosemblatt, 187–209. Chapel Hill: University of North Carolina Press, 2003.
Mireles Gavito, Sofía. "Construcción del Ferrocarril Panamericano." *La Voz del Norte. Periódico cultural de Sinaloa*, September 30, 2012. https://www.lavozdelnorte.com.mx/2012/09/30/construccion-del-ferrocarril-panamericano/#:~:text=Despu%C3%A9s%20se%20empieza%20a%20construir,1%20de%20noviembre%20de%201904.
Misawa Saito, Katsuhito. "La colonia Enomoto de Chiapas. Estrategia expansionista y proyectos migratorios japoneses a fines del siglo XIX: el caso de México." MA thesis, Universidad Nacional Autónoma de México/Facultad de Filosofía y Letras, 1982.
Molina Pérez, Valente. "Impacto económico y social del Ferrocarril Panamericano en la región de Tonalá en el siglo XX." *Revista Pueblos y Fronteras Digital* 11, no. 21 (January–June 2016): 67–91. https://doi.org/10.22201/cimsur.18704115e.2016.21.9.
Monteón González, Humberto, and José Luis Trueba Lara. *Chinos y antichinos en México. Documentos para su estudio*. Guadalajara: Gobierno del Estado de Jalisco, 1988.
Moreno, Alejandra. "Cambios en los patrones de urbanización en México, 1810–1910." *Historia Mexicana* 22, no. 2 (1972): 160–87.
Moreno Corral, Marco Arturo. *Odisea 1874 o El primer viaje internacional de científicos mexicanos*. Mexico City: FCE, 2003.

Ochoa Rodríguez, Héctor Porfirio. "Manzanillo: el intrincado despertar de un puerto." In *Los puertos noroccidentales de México*, edited by Jaime Olveda and Juan Carlos Reyes, 113–19. Zapopan: El Colegio de Jalisco, 1994.

Oliver, Jason. *Chino: Anti-Chinese Racism in Mexico, 1880–1940*. Champaign: University of Illinois Press, 2017.

Oropeza, Déborah. *La migración asiática en el virreinato de la Nueva España: un proceso de globalización, 1565–1700*. Mexico City: El Colegio de México, 2020.

Ota Mishima, María Elena, ed. *Destino México: Un estudio de las migraciones asiáticas a México, siglos XIX y XX*. Mexico City: El Colegio de México, 1997.

Ota Mishima, María Elena. *Siete migraciones japonesas en México, 1890–1978*. Mexico City: El Colegio de México, 1982.

Palma Mora, Mónica. "De la simpatía a la antipatía. La actitud oficial ante la inmigración, 1908–1990." *Historias* 56 (2003): 63–76, https://www.revistas.inah.gob.mx/index.php/historias/article/view/12952.

Parejo, Juan. "Una versión inédita de la vuelta al mundo de Magallanes y Elcano." *Diario de Sevilla*, November 7, 2019. https://www.diariodesevilla.es/sevilla/version-inedita-vuelta-mundo-Magallanes-Elcano_0_1407459578.html.

Patterson, Wayne. *The Korean Frontier in America: Immigration to Hawaii, 1896–1910*. Honolulu: University of Hawai'i Press, 1988.

Paulo, María. *Origen de Salina Cruz*. Oaxaca: Familia Moreno Paulo, 1977.

Peddie Robson, Francis David. "La colonia japonesa de México y la Segunda Guerra Mundial." MA thesis, Facultad de Filosofía y Letras/UNAM, 2005.

Peña Delgado, Grace. *Making the Chinese Mexican: Global Migration, Localism, and Exclusion in the U.S.–Mexico Borderlands*. Stanford, CA: Stanford University Press, 2012.

Perez, Louis G. *Japan Comes of Age: Mutsu Munemitsu and the Revision of the Unequal Treaties*. London: Associated University Presses, 1999.

Pinzón Ríos, Guadalupe. "Acciones y reacciones en los puertos del Mar del Sur. Desarrollo portuario del Pacífico novohispano a partir de sus políticas defensivas, 1713–1789." PhD diss., Facultad de Filosofía y Letras, UNAM, 2008.

Pletcher, David M. *The Diplomacy of Involvement: American Economic Expansion across the Pacific, 1784–1900*. Columbia: University of Missouri Press, 2001.

Pletcher, David M. *Rails, Mines, and Progress: Seven American Promoters in Mexico, 1867–1911*. Port Washington: Kennikat, 1972.

Pratt, Mary Louise. *Imperial Eyes: Travel Writing and Transculturation*. New York: Routledge, 1992.

Pratt, Mary Louise. "Scratches on the Face of the Country; or, What Mr. Barrow Saw in the Land of the Bushmen." In *Defining Travel: Diverse Visions*, edited by Susan Roberson, 132—52. Jackson: University Press of Mississippi, 2007.

Price, John. "Relocating Yuquot: The Indigenous Pacific and Transpacific Migrations." *BC Studies* 204 (Winter 2019–2020): 21–44.

Puig Llano, Juan. *Entre el río Perla y el Nazas: La China decimonónica y sus braceros emigrantes, la colonia china de Torreón y la matanza de 1911*. Mexico City: Conaculta, 1992.

Quero Morales, José. "El derecho sanitario mexicano." *Revista de la Facultad de Derecho de la UNAM* (January–March 1963): 141–76.

Quirarte, Vicente. *Republicanos en otro imperio. Viajeros mexicanos a Nueva York (1830–1895)*. Mexico City: UNAM, 2009.
Ratz, Konrad. *Maximiliano de Habsburgo*. Mexico City: Planeta, 2002.
Ratz, Konrad. *Tras las huellas de un desconocido: nuevos datos y aspectos de Maximiliano de Habsburgo*. Mexico City: Conaculta/Siglo XXI, 2008.
Reid, Anthony. *The Chinese Diaspora in the Pacific*. Burlington, VT: Ashgate Variorum, 2008.
Relación de la visita oficial a la zona de la colonia Enomoto de la Chiapas, sur de México. Mexico City: 1958.
Rénique, Gerardo. "Race, Region, and Nation: Sonora's Anti-Chinese Racism and Mexico's Postrevolutionary Nationalism, 1920s–1930s." In *Race and Nation in Modern Latin America*, edited by Nancy P. Appelbaum, Anne S. Macpherson, and Karin Alejandra Rosemblatt, 211–36. Chapel Hill: University of North Carolina Press, 2003.
Riguzzi, Paolo. "Los caminos del atraso: tecnología, instituciones e inversión en los ferrocarriles mexicanos, 1850–1900." In *Ferrocarriles y vida económica en México, 1850–1950. Del surgimiento tardío al decaimiento precoz*, edited by Sandra Kuntz and Paolo Riguzzi, 31–97. Mexico City: El Colegio Mexiquense/UAM-X/FNM, 1996.
Riguzzi, Paolo. "Propiedad, propietarios y recursos nacionales en los ferrocarriles mexicanos, 1870–1905." In *Memorias del Tercer Encuentro de Investigadores del Ferrocarril*. Puebla: Museo Nacional de los Ferrocarriles, 1996.
Roberson, Susan. *Defining Travel: Diverse Visions*. Jackson: University Press of Mississippi, 2007.
Rodríguez, Nemesio J. *Istmo de Tehuantepec: de lo regional a la globalización o apuntes para pensar un quehacer*. Oaxaca: Gobierno del Estado de Oaxaca, 2003.
Román Alarcón, Rigoberto Arturo. *El comercio en Sinaloa, siglo XIX*. Culiacán: DIFOCUR/FOECA/CONACULTA, 1998.
Román Alarcón, Rigoberto Arturo. *La economía del sur de Sinaloa, 1910–1950*. Mazatlan: Instituto Municipal de Cultura, Turismo y Arte de Mazatlán/DIFOCUR, 2006.
Romero Castilla, Alfredo. "Huellas del paso de los inmigrantes coreanos en tierras de Yucatán y su dispersión por el territorio mexicano." In *Destino México: Un estudio de las migraciones asiáticas a México, siglos XIX y XX*, edited by María Elena Ota Mishima, 123–66. Mexico City: El Colegio de México, 1997.
Romero Gil, Juan Manuel. *La minería en el noroeste de México: utopía y realidad 1850–1910*. Mexico City: Universidad de Sonora/Plaza y Valdés, 2001.
Ronzón León, José Agustín. "Modernidad, sanidad y nacionalismo en el México porfirista. Una mirada historiográfica a través del código sanitario de 1894." *Tzintzun. Revista de Estudios Históricos* 75 (January–June 2022): 63–88.
Rosemblatt, Karin Alejandra. *The Science and Politics of Race in Mexico and the United States, 1910–1950*. Chapel Hill: University of North Carolina Press, 2018.
Samson, Jane. *British Imperial Strategies in the Pacific, 1750–1900*. Aldershot: Ashgate, 2003.
Schaefer, Timo. *Liberalism as Utopia: The Rise and Fall of Legal Rule in Post-Colonial Mexico, 1820–1900*. New York: Cambridge University Press, 2017.
Schell, William Jr. "Silver Symbiosis: Reorienting Mexican Economic History." *Hispanic American Historical Review* 81, no. 1 (2001): 89–133.

Schiavon, Jorge A., Daniela Spenser, and Mario Vázquez Olivera, eds. *En busca de una nación soberana. Relaciones internacionales de México, siglos XIX y XX*. Mexico City: SRE/CIDE, 2006.

Schiavone Camacho, Julia María. *Chinese Mexicans: Transpacific Migration and the Search for a Homeland, 1910–1960*. Chapel Hill: University of North Carolina Press, 2012.

Schurz, William Lytle. *El galeón de Manila*. Madrid: Ediciones de Cultura Hispánica, 1992.

Seijas, Tatiana. "Asian Migrations to Latin America in the Pacific World, 16th–19th Centuries." *History Compass* 14 (2016): 573–81.

Seijas, Tatiana. *Asian Slaves in Colonial Mexico: From Chinos to Indians*. New York: Cambridge University Press, 2014.

Serrano Álvarez, Pablo. "Comentario." In *Los puertos noroccidentales de México*, edited by Jaime Olveda and Juan Carlos Reyes, 66–70. Zapopan: El Colegio de Jalisco, 1994.

Shicheng, Xu. "Algunas reflexiones sobre el desarrollo de las relaciones sino-mexicanas." *Cuadernos Americanos* 121 (July–September 2007): 171–86.

Shicheng, Xu. "Los chinos a lo largo de la historia de México." In *China y México: Implicaciones de una nueva relación*, edited by Enrique Dussel Peters and Yolanda Trápaga Delfín, 51–68. Mexico City: La Jornada Ediciones, 2007.

Shipping, Technology, and Imperialism: Papers Presented to the Third British–Dutch Maritime History Conference. Brookfield, VT: Ashgate, 1996.

Sierra de la Calle, Blas. "La expedición de Legazpi-Urdaneta (1564–1565). El tornaviaje y sus frutos." In *Instituto de Historia y Cultura Naval. Cuaderno monográfico nº 58*, 129–67. Madrid: Ministerio de Defensa, 2009. https://armada.defensa.gob.es/archivo/mardigitalrevistas/cuadernosihcn/58cuaderno/cap05.pdf.

Silva Castañeda, Sergio, and Graciela Márquez Colín. *Matías Romero and the Craft of Diplomacy: 1837–1898*. Mexico City: SRE, Instituto Matías Romero, 2018.

Sinn, Elisabeth. *Pacific Crossing: California Gold, Chinese Migration, and the Making of Hong Kong*. Hong Kong: Hong Kong University Press, 2013.

Skirius, John. "Railroad, Oil and Other Foreign Interests in the Mexican Revolution, 1911–1914." *Journal of Latin American Studies* 35, no. 1 (February 2003): 25–51.

Sodi Álvarez, Enrique. *Istmo de Tehuantepec*. Mexico City: Talleres Gráficos de la Nación, 1967.

Sohn, Rebecca, and Doris Elin Urrutia. "Astronomical Unit: How Far Away Is the Sun?" Space.com. Last modified November 1, 2023. https://www.space.com/17081-how-far-is-earth-from-the-sun.html.

Spate, O. H. K. *Monopolists and Freebooters*. London: Croom Helm, 1983.

Spate, O. H. K. *Paradise Found and Lost*. London: Routledge, 1988.

Spate, O. H. K. *The Spanish Lake*. London: Croom Helm, 1979.

Starr, Frederick. *Mexico and the United States: A Story of Revolution, Intervention, and War*. Chicago: University of Chicago Press, 1914.

Steele, M. William. *Alternative Narratives in Modern Japanese History*. London: Routledge, 2003.

Stepan, Nancy L. *"The Hour of Eugenics": Race, Gender, and Nation in Latin America*. Ithaca, NY: Cornell University Press, 1991.

Sue, Christina A. "Is Mexico Beyond Mestizaje? Blackness, Race Mixture, and Discrimination." *Latin American and Caribbean Ethnic Studies* 18, no. 1 (January 2023): 47–74.

Tate, E. Mowbray. *Transpacific Steam: The Story of Steam Navigation from the Pacific Coast of North America to the Far East and the Antipodes, 1867–1941.* New York: Cornwall Books, 1986.

Telles, Edward E. *Pigmentocracies: Ethnicity, Race and Color in Latin America.* Chapel Hill: University of North Carolina Press, 2014.

Tenorio Trillo, Mauricio. "1910 Mexico City: Space and Nation in the City of the Centenario." *Journal of Latin American Studies* 28 (February 1996): 75–104.

Tenorio Trillo, Mauricio, and Aurora Gómez Galvarriato. *El Porfiriato.* Mexico City: FCE, 2006.

TheShipsList. "Colorado." Accessed October 4, 2023. https://www.theshipslist.com/ships/descriptions/panamafleet.shtml.

Toussaint Ribot, Mónica. "Los negocios de un diplomático: Matías Romero en Chiapas." *Latinoamérica. Revista de estudios Latinoamericanos* 55 (2012): 129–57.

Tovar González, María Elena. "Extranjeros en el Soconusco." *Revista de Humanidades, ITESM* 8 (2000): 35–36.

Trejo Barajas, Dení. "Del Mar Caribe al Mar del Sur. Comercio marítimo por el Pacífico mexicano durante las guerras de independencia." In *Entre la tradición y la modernidad. Estudios sobre la independencia*, edited by Moisés Guzmán, 353–80. Morelia: UMSNH, 2006.

Trueba Lara, José Luis. *Los chinos en Sonora: una historia olvidada.* Hermosillo: Universidad de Sonora, 1990.

Trujillo Bolio, Mario. *El Golfo de México en la centurión decimonónica. Entornos geográficos, formación portuaria y configuración marítima.* Mexico City: Porrúa/CIESAS/Cámara de Diputados LIX Legislatura, 2005.

Uscanga, Carlos. "Hacia una contextualización histórica de las relaciones diplomáticas de México y Japón." *Revista Mexicana de Política Exterior* 86 (June 2009): 67–89.

Valdés Lakowsky, Vera. "Cambios en las relaciones transpacíficas: del *Hispanis Mare Pacificum* al Océano Pacífico como vía de comunicación internacional." *Revista de Estudios de Asia y África* 53, no. 1 (1985): 58–81.

Valdés Lakowsky, Vera. *De las minas al mar. Historia de la plata mexicana en Asia, 1565–1834.* Mexico City: FCE, 1987.

Valdés Lakowsky, Vera. "Entre proyecto e improvisación: la plata mexicana en el Pacífico." In *La inserción de México en la cuenca del Pacífico*, Vol. 1, edited by Alejandro Álvarez and John Borrego, 125–35. Mexico City: UNAM, 1990.

Valdés Lakowsky, Vera. "Ignacio Mariscal." In *Cancilleres de México*, vol. 1, 579–83. Mexico City: SRE, 1992.

Valdés Lakowsky, Vera. "México y China: del galeón de Manila al primer tratado de 1899." *Estudios de Historia Moderna y Contemporánea de México* 9 (1983): 9–19.

Valdés Lakowsky, Vera. *Vinculaciones sino-mexicanas: albores y testimonios, 1874–1899.* Mexico City: UNAM, 1981.

Vargas, Rosa Elvira. "Embajador apremia a auxiliar a Japón." *La Jornada*, April 6, 2011.

Vega y Ortega Baez, Rodrigo Antonio, and Gustavo Enrique Flores Herrera. "Un funcionario experto. La producción geográfica del ingeniero Manuel Fernández Leal, 1877–1911." *Investigaciones Geográficas* 109 (December 2022). https://doi.org/10.14350/rig.60581.

Velasco Ávila, Cuauhtémoc, Eduardo Flores Clair, Alma Laura Parra Campos, and Edgar Omar Gutiérrez López. *Estado y minería en México 1767-1910*. Mexico City: FCE/ Secretaría de energía, minas e industria paraestatal, 1988.

Velázquez Becerril, César Arturo. "Intelectuales y poder en el porfiriato: una aproximación al grupo de los científicos, 1892-1911." *Revista Fuentes humanísticas* 22, no. 41 (December 2010): 7-23.

Vertovec, Steven. "Migration and Other Modes of Transnationalism: Towards Conceptual Cross-Fertilization." *International Migration Review* 37, no. 3 (Fall 2003): 641-65.

Villegas Revueltas, Silvestre. "La deuda inglesa: el componente de la relación anglo-mexicana." In *En busca de una nación soberana. Relaciones internacionales de México, siglos XIX y XX*, edited by Jorge A. Schiavon, Daniela Spenser, and Mario Vázquez Olivera, 157-99. Mexico City: SRE, 2006.

Wade, Peter. *Race and Ethnicity in Latin America*. New York: Pluto Press, 2010.

Walworth, Arthur. *Black Ships Off Japan: The Story of Commodore Perry's Expedition*. New York: Alfred A. Knopf, 1946.

Washbrook, Sarah. *Producing Modernity in Mexico: Labour, Race, and the State in Chiapas, 1876-1914*. London: The British Academy, 2012.

Watanabe, Chizuko. "The Japanese Immigrant Community in Mexico: Its History and Present." MA thesis, California State University, 1983.

Weiner, Richard. *Race, Nation, and Market: Economic Culture in Porfirian Mexico*. Tucson: University of Arizona Press, 2004.

Wickberg, Edgar. *The Chinese in Philippine Life, 1850-98*. New Haven, CT: Yale University Press, 1965.

Williams, Michael. *Returning Home with Glory: Chinese Villagers around the Pacific, 1849 to 1949*. Hong Kong: Hong Kong University Press, 2018.

Yankelevich, Pablo. "Revolución e inmigración en México (1908-1940)." *Anuario de la Revista Digital de la Escuela de Historia* 24, no. 3 (2011-2012): 39-71. https://doi.org/10.35305/aeh.v0i24.97.

Yankelevich, Pablo, and Paola Chenillo Alazraki. "El Archivo Histórico del Instituto Nacional de Migración." *Desacatos* 26 (January-April 2008): 25-42, https://desacatos.ciesas.edu.mx/index.php/Desacatos/article/view/535/401.

Young, Dana B. "The Voyage of the Kanrin Maru to San Francisco, 1860." *California History* 61, no. 4 (Winter 1983): 264-75.

Young, Elliott. *Alien Nation: Chinese Migration in the Americas from the Coolie Era through World War II*. Chapel Hill: University of North Carolina Press, 2014.

Yu, Henry. "Unbound Space: Migration, Aspiration, and the Making of Time in the Cantonese Pacific." In *Pacific Futures: Past and Present*, edited by Warwick Anderson, Miranda Johnson, and Barbara Brookes, 178-204. Honolulu: University of Hawai'i Press, 2018.

Yuste, Carmen. *El comercio de la Nueva España con Filipinas, 1590-1785*. Mexico City: INAH, 1984.

Zamora, Pedro. "Según reportes de 1908 a 1914 de la Oficina de Inteligencia Naval estadounidense." *Proceso*, May 15, 1999.

Zavala, Silvio. *Apuntes de historia nacional, 1808-1974*. Mexico City: FCE, 1990.

Zoraida Vázquez, Josefina, ed. *Interpretaciones del periodo de Reforma y Segundo Imperio*. Mexico City: Patria, 2007.

www.ingramcontent.com/pod-product-compliance
Ingram Content Group UK Ltd.
Pitfield, Milton Keynes, MK11 3LW, UK
UKHW031854161224
452368UK00004B/96